普通高等教育“十二五”规划教材

Mining of Non-ferrous Metal Deposits

有色金属矿床开采

主　编　占丰林　叶　萍
副主编　凌征华　俞　惠

北　京
冶金工业出版社
2011

内 容 提 要

本书为冶金类院校英语专业和采矿工程专业开设的有色金属方面特色课程所用的教材之一。全书共分6章，主要内容包括：采矿和有色金属矿床简介、炸药与爆破、有色金属矿山运输、有色金属矿山通风、岩石力学以及有色金属矿床采矿方法等。为便于教学，每章前附有本章的中文内容摘要，每章后附有本章的专业术语词汇表及复习思考题。

本书可作为冶金类院校英语专业本科生的泛读教材、采矿工程专业本科生的专业英语教材和采矿概论双语课程的教材，也可作为采矿工程专业人员出国考察访问、进修或攻读学位前的培训或自学教材，还可供广大采矿工程师及采矿工程技术人员学习参考。

图书在版编目(CIP)数据

有色金属矿床开采 = Mining of Non-ferrous Metal Deposits：英文/占丰林，叶萍主编. —北京：冶金工业出版社，2011.5

普通高等教育"十二五"规划教材

ISBN 978-7-5024-5463-0

Ⅰ.①有… Ⅱ.①占… ②叶… Ⅲ.①有色金属矿床—金属矿开采—高等学校—教材—英文 Ⅳ.①TD862

中国版本图书馆 CIP 数据核字(2011)第055280号

出 版 人 曹胜利
地　　址 北京北河沿大街嵩祝院北巷39号，邮编100009
电　　话 (010)64027926 电子信箱 yjcbs@cnmip.com.cn
责任编辑 廖 丹 美术编辑 彭子赫 版式设计 孙跃红
责任校对 王永欣 责任印制 张祺鑫
ISBN 978-7-5024-5463-0
北京印刷一厂印刷；冶金工业出版社发行；各地新华书店经销
2011年5月第1版，2011年5月第1次印刷
787mm×1092mm 1/16；15.5印张；371千字；230页
32.00元

冶金工业出版社发行部 电话:(010)64044283 传真:(010)64027893
冶金书店 地址:北京东四西大街46号(100010) 电话:(010)65289081(兼传真)

前　言

随着我国市场经济的发展壮大，行业分工越来越细、越来越综合，对专门人才、特色人才和复合应用型人才的需求日益旺盛，这对我国高等教育的人才培养也产生了巨大的影响。2007年，教育部高等学校外语专业教学指导委员会评估组专家对一些大学的英语专业本科教学工作提出了“走特色办学之路”的建议。因此，高校专业特色化是当前人才培养的方向。

有色金属采矿、选矿和冶金是冶金类院校的强势特色专业。因此，此类高校的英语专业应具有有色冶金特色。特色专业需要特色系列教材，目前市场上正式出版的有色金属采矿、有色金属选矿和有色金属冶金方面的英文教材比较少。同时冶金类院校采矿工程专业使用的专业英语教材的专业知识不够深入系统。本书是冶金类院校英语专业和采矿工程专业开设的有色金属方面特色课程所用的教材之一，旨在使这些院校英语专业和采矿工程专业的学生通过专业英语的学习掌握采矿工程专业知识，使英语专业和采矿工程专业的学生不仅能够用英语进行日常会话、翻译，还可以熟练地使用英语进行采矿工程专业方面的听、说、读、写、译，这也是当今冶金采矿行业最稀缺的复合应用型人才之一。

本书可作为冶金类院校英语专业本科生的泛读教材、采矿工程专业本科生的专业英语教材和采矿概论双语课程的教材，也可作为采矿工程专业人员出国考察访问、进修或攻读学位前的培训或自学教材，还可供广大采矿工程师及采矿工程技术人员学习参考。在本书实际使用过程中，使用者可根据使用目的的不同和学时的多少进行选择性的讲解、学习或阅读。

全书共分6章，主要内容包括：采矿和有色金属矿床简介、炸药和爆破、有色金属矿山运输、有色金属矿山通风、岩石力学以及有色金属矿床采矿方法等。本书由占丰林教授、叶萍副教授担任主编，凌征华教授、俞惠副教授担任副主编。其中第1、3、5、6章由占丰林编写，第2章由叶萍编写，4.3~4.6节由凌征华编写，4.1～4.2节由俞惠编写，全书由占丰林统稿。

本书在编写过程中参考了国内外相关书籍、论文或网页，在此对这些参考资料的作者表示诚挚的感谢。由于编者水平有限，书中难免有疏漏之处，敬请广大读者批评指正。

编　者

2011年元月

CONTENTS

目　　录

1 Introduction to Mining and Non-ferrous Metal Deposits

本章介绍一些采矿专业术语、有色金属矿床概念、采矿与人类文明、采矿对环境的影响、矿床评价与资源保护以及有用矿物回收方法，最后阐述中国有色金属工业的现状与未来。

1.1 Mining's Role

1.1.1 Minerals, Rocks and Ore

In the year 1971, the number of known elements amounted to 104. Fifteen of these have been made only in the laboratory, others may have persisted from the primitive atmosphere, but, by and large, elements originated from magmas or igneous rocks of the outer rocky shell of the earth. Only eight elements constitute 98% by weight of the earth's crust. These are oxygen, 47%; silicon, 28%; aluminum, 8%; iron, 5%; and sodium, magnesium, potassium and calcium, less than 4% each. These common elements and the other less common ones are the building blocks of minerals, of which there are over 2000 varieties.

According to Webster: "A mineral is an inorganic substance occurring naturally in the earth and having a consistent and distinctive set of physical properties and a composition that can be expressed by a chemical formula. The term is sometimes applied to organic substances, such as coal." Thus, minerals are precise combinations of elements. Rocks, as distinct from minerals, are composed of assemblages of minerals.

When minerals are found in sufficient concentration to warrant extraction by mining, the mineralized area is considered an ore deposit. The definition of ore is mineral that can be extracted from the ground at a profit. The economic connotation is implicit in the word ore.

Since most of the useful elements compose such a small percentage of the earth's crust, the occurrence of ore deposits as we know them would not have transpired had not geologic processes concentrated the elements.

1.1.2 Professional Terms of Mining

Extensive coverage of the many descriptive terms used in mining may be found in a good mining glossary, such as is available from the Superintendent of Documents, but for convenience a number of the more common definitions are given here.

Mining may be defined, as by A. B. Cummins, as the act, process or work of extracting minerals or coal from their natural environment and transporting them to the point of processing or use. Mining techniques are applied to extracting metallic minerals, such as ores of gold, copper, lead or zinc; to fuels, such as coal, anthracite, lignite and tar sands; and to nonmetallic minerals, such as limestone, sand and gravel, clay and stone, sulfur and salt. These are just a few of the many minerals extracted by mining processes.

A mine, therefore, is an excavation made in the earth for the purpose of extracting minerals. Such excavations may be at the surface or underground, or both surface and underground methods may be employed.

The selection of mining methods is made to effect the most economic recovery of the minerals (see Section 6.1). Formerly, the type of deposit and the physical characteristics of the deposit and the enclosing rocks were the dominant factors in selecting the mining method. Today, preservation of the environment can be equally important in determining the method.

Most base metal deposits are composed of several valuable minerals mixed with waste minerals called gangue. In mining, the objective is to extract as much of the valuable mineral as possible and leave behind the waste. The material coming from the mine, called ore, is naturally a combination of valuable minerals and gangue.

The ore, in the case of metallic deposits, usually is processed at the mine to concentrate the valuable mineral for shipment to the smelter or other refining facility. The gangue is discarded.

Nonmetallic deposits and fuels, unlike metalliferous deposits, contain more of the valuable mineral or element than of waste. Nonmetallic minerals, also called industrial minerals, are prepared at the mine into a finished product, or the prepared mineral may be shipped elsewhere for additional treatment before being marketed.

Coal usually is washed at the mine to remove to the maximum extent possible slate and other impurities, and then shipped to the power plant, coke ovens or other consumer. A modern development is the construction of power plants at the mine, which have become known as "mine mouth power plants."

There are several types of surface mines. In an open-pit mine, an excavation open to the surface is dug to extract metallic ores (see Section 6.3). In the typical open pit, digging progresses in terraces. In a strip mine, the overlying surface material (called spoil, overburden or waste) is removed to expose a coal bed for digging. A quarry is an excavation for extracting stone, sand or gravel. In a placer mine, the valuable material is found on the bedrock beneath gravel in stream beds, fossil stream beds or flood plains and is recovered by panning, hydraulicking or dredging. Seawater processing, wherein salt or magnesium is extracted by evaporation or chemical processing, although not strictly mining, falls within the province of the minerals industry.

In underground mines, diverse extraction methods are employed (see Section 6.2). Some of these are block caving, top slicing, sublevel caving, longhole stoping, shrinkage stoping, open stoping, square setting and many others. The reason there are many methods is that mineral deposits have different shapes and the mineral and host rock of each ore body have differing physical characteristics, all having a bearing on the choice of mining method.

Coal deposits are mined underground by several methods but not so many as for mineral deposits because most economic coal deposits are similar, being tabular, relatively flat-lying beds of great areal extent. The chief methods of extracting coal underground are by room-and-pillar and longwall mining.

Another important underground mining method is solution mining. In a solution mine, a liquid is pumped into the ground through a borehole to liquefy or dissolve a deposit, which then is pumped to the surface. Examples are sulfur and salt mining. Copper and uranium also are extracted by utilizing chemical solutions to dissolve them from ores in their natural environment.

In a hydraulic mine, jets of water break down the ore-bearing gravel so that it can be pumped to the concentrating plant. Placer deposits of tin or gold are mined in this manner. This method also is utilized for extracting the mineral gilsonite underground and it has been tested on a small scale for extracting underground coal and phosphate beds.

Ocean or offshore mining covers the extraction of minerals resting on the ocean floor or in the strata underlying the oceans. Dredging of minerals in water up to nearly 61m in depth on continental shelf areas has been practiced for many years in different parts of the world. There also are mines for coal, iron and other minerals that have been extended from shore out under the ocean floor. The ocean floor and substrata hold great potential as sources for future mineral production.

Some other terms that describe underground mines refer to the means of entry and egress for men, equipment and mineral. For example, a shaft mine has a vertical or inclined opening, an adit mine is a metal mine with a horizontal entry (in coal, a drift mine). A slope mine, in coal, is served by an inclined opening—the slope.

The foregoing descriptions regarding surface and underground mines are terms used by mining men, rather than strict definitions. For example, although surface mines on deposits of metallic minerals always are called open-pit mines, it would not be wrong to call a surface mine on a nonmetallic deposit an open-pit mine. Although strip mining generally applies to coal, some surface uranium mines are called strip mines. The word placer is a geologic description of a deposit which has come to be used to describe a form of mining. While most placers are surface mines, some are worked through adits and shafts. The term quarry, generally regarded as a surface mine, may be applied to an underground excavation. It has been applied especially to mining of building stone.

Mines also are divided into three broad commodity categories: metallic, nonmetallic and coal. This grouping is convenient because of generalized similarities within each category with respect to geology, mining techniques, processing and marketing. Metallic mines are usually divided into two categories: ferrous and non-ferrous metal mines. Sometimes, mines are divided into two groups: coal and noncoal.

Mining techniques also are applied to the removal of earth or rocks for military or civil purposes. In military or civil mining, the objective usually is the opening. Another major difference between mining for these purposes and mineral mining is in latitude for site selection. Whereas mining for minerals must be on the location of the ore deposit, selection of a site where mining is performed for military or civil purposes is governed by numerous factors. These factors can be rigidly defined by considerations other than ease of mining. Another important difference between military and civil mining and mineral mining is that time usually is important in the two former cases, either through necessity or economics. Time is important for economic reasons in mineral mining during the development stage, but when the mine is producing, which often is for a period of many years, time ceases to be of great importance. Regardless of the objective of mining work, there is considerable similarity in methods and machines.

A mining engineer specializes in the working of mines. In this profession, a knowledge of many branches of engineering and science, such as civil, mechanical, electrical and metallurgical engineering and geology and chemistry, is needed. A person may become a mining engineer by education followed by experience in the field. The mining engineer can devote a portion of his career or all of it to a specialized type of mining or to a phase of mining activity.

The mining man cannot ignore the marketing functions of his organization. For example, a company with a large marketing responsibility in metals will direct its geologists to search only for big deposits. In a certain nonmetallic operation, the requirements of a new market may necessitate more selective mining. In metallic, nonmetallic and coal mining, the market price of the product is the watchdog of efficiency.

The schematic diagram (Fig. 1-1) depicts the scope of mining activities from the exploration for a mineral deposit through the finding, proving, developing, mining and processing to marketing the products of the mining enterprise. Each step requires planning and design as well as execution. Many specialists collaborate to achieve a successful mining venture. The mining engineer primarily is involved in the Mining Proper phase and in the development work which precedes mining from the standpoints of planning, design and execution. He may also supervise or manage a phase or become an executive of a mining company with a broad range of responsibility. Since the mining of a mineral deposit usually endures for many years, the planning and engineering are continuing functions which must predict and be responsive to changes in character of the ore deposit, technologic developments and economics.

1.1.3 Mining and Civilization of Mankind

Nature provides only six sources of new wealth. These may be enumerated as the fruits of the following endeavors: parentage, hunting and fishing, forestry, agriculture, husbandry and mining.

From prehistory to the present, mining has provided materials for fuel, shelter and the acquisition of food. The great impact of the products of mining on Man is evident because anthropologists designate steps in Man's culture by them; Paleolithic (Old Stone Age), Neolithic (New Stone

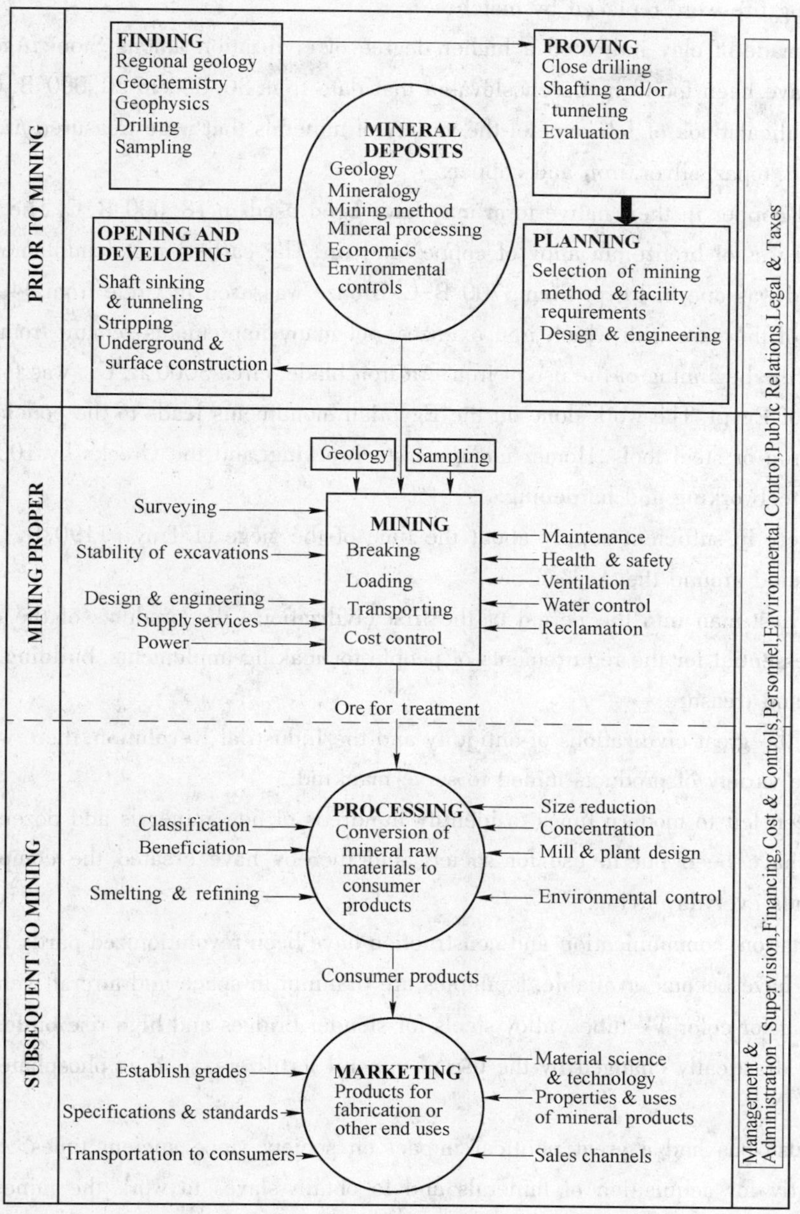

Fig. 1-1 Schematic diagram depicting the scope of mining activities from exploration to final marketing of the output

Age), Bronze Age (4000—1800 B. C. in the Orient; 2000—1000 B. C. in Europe) and Iron Age (following the Bronze Age). Today, the standard of living of the peoples of the world are compared on the basis of per capita consumption of the various metals.

Stone Age men utilized flint for thousands of years to fashion tools for the chase and to make fire. Mines for flint have been found at Obourg and Spiennes in Belgium and in numerous other localities that were worked in the period 3000—2000 B. C. It was not until 1835 A. D. that flint and steel for making fire were replaced by matches.

Utensils made of clay indicated a higher degree of civilization among Stone Age men. Fired clay objects have been found in Czechoslovakia that date from 30,000 to 20,000 B. C.

In the Biblical Book of Job some of the variety of minerals that were treasured are mentioned. They were gold, topaz, silver, iron and copper.

Gold and copper in their native form may have been used in 18,000 B. C. The first common usage of metal was of bronze, an alloy of copper and tin. The earliest metal implement found that has been dated was one in Egypt from 3700 B. C. Bronze was used in Crete from 3400 B. C.

Because of the ease with which iron oxidizes, not many implements remain from antiquity by which to date the beginning of the use of iron. An iron blade, circa 3000 B. C., was found in one of the pyramids of Egypt. The work done on the Egyptian monuments leads to the conclusion that the workers had iron or steel tools. Homer mentions steelworking, and the Greeks by 1000 B. C. were advanced in steelworking and hardening.

Metals were in sufficient supply about the time of the siege of Troy (1190 B. C.) that they were being traded around the Aegean Sea.

From earliest man into the period of the first civilizations, the products of the earth won by mining were essential for the requirements of people for making implements, building, trade, jewelry, cosmetics and treasure.

Between the great civilizations of antiquity and the Industrial Revolution, there were not many changes in the variety of products mined to serve mankind.

It has been left to modern times to identify hundreds of new minerals and dozens of new elements. These have been put to use for society and thereby have created the complex hardware which surrounds everyday living.

Transportation, communication and construction have been revolutionized partly because of the materials that have become available. Examples are titanium in space and aircraft vehicles, europium and yttrium for color TV tubes, alloy steels for slender bridges and high rise buildings. Agriculture has been drastically changed by the use of mineral fertilizers, such as phosphates, potash and nitrates.

Mining also has had a great political impact on society since ancient times. Ancient rulers made war partly for acquisition of minerals and to obtain slaves to work the mines. Cortes and Pizarro came to the New World for booty. The opening of the American West was accelerated by the discovery of gold in California. Japan invaded Manchuria prior to World War II for the acquisition of coal and iron. Underlying the politics of Russia and the United States in the Mideast are the rich oil reserves.

1.1.4 Advancement of Mining Technology

Primitive miners utilized their hands and implements of wood, bone, stone and, later, of metal with which to dig and extract minerals. With the advent of a social system, mining probably became more organized, utilizing slaves laboring under supervisors who no doubt had production goals. In the ancient societies, implements were improved, the wedge and mallet were devised, baskets were made for carrying rock or removing water, ladders and windlasses aided transportation, illumination with candles or other lamps was introduced.

Underground mines were dug to depths of several hundred meter. For example, ancient Egyptian emerald workings on the Red Sea were 243.8m deep and extensive enough to employ 400 workers at a time. Some Roman mines in Spain were 198.12m deep.

The legend of Jason and the Golden Fleece is a reference to the first recorded placer mining of fine gold (sheep' s wool placed in streams caught the fine gold).

Mining was common in ancient times around the Mediterranean. Ancients such as Herodotus and Aristotle mentioned mining but notable treatises on mining were not forthcoming until the 16th century. The most famous of these is the De Re Metallica by Georgius Agricola, published in 1556. The only English translation is by Herbert C. Hoover and his wife, Lou Henry Hoover, published in 1912.

In the Western Hemisphere, the Indians, especially the Incas, mined silver and gold before the arrival of the Spaniards. The first recorded mining in the New World was copper in Cuba, by Spaniards in 1524. The settlers of New England did a little mining. Two Frenchmen began mining lead in Missouri at Mine la Motte in 1718.

The Industrial Revolution created a demand for metals which intensified the search for minerals and accelerated new mine development in the 17th and 18th centuries. The first important change in mining practice was the introduction of black powder for blasting rock in 1627 in Hungary. Soon afterward, black powder was at work in the tin mines of Cornwall. In 1718, the Cornish tin mines were being dewatered by pumping. This was a great advance for it permitted following the veins to greater depths.

The steam engine and the air compressor amplified enormously over human muscle the energy that could be applied to mining. In the late 19th century, the air-driven rock drill replaced hand drilling. The invention of electricity gave great impetus to mechanization and made applications of machinery more flexible.

The introduction of the steam shovel to the open-pit mines just prior to World War I was another milestone in increasing productivity.

A panorama of the mineral industry in the decade of the 1960s illustrates the degree of technologic progress since 1920.

In underground coal mines, coal is cut from the seam by continuous mining machines of various types without recourse to drilling and blasting and the coal is moved continuously from the mining face to the surface by belt conveyors. In the strip mines, probably the largest machines in the world are being used to remove the overburden and dig the coal. The largest dragline in 1970 has a bucket capacity of $168m^3$ and the largest shovel, $138m^3$.

In underground gold mining in South Africa, there are men drilling and blasting $3,219m^3$ beneath the surface extracting ore that averages less than 28g of gold in each metric ton of rock. At some underground mines, such as the White Pine copper mine in Michigan and in the lead mines of Missouri, it is seldom that a miner works on foot. Self-propelled drill jumbos with hydraulically operated drill arms move up to the working face and quickly drill 3.7m holes. These are charged pneumatically with blasting agents from a special vehicle. The blasted rock is dug by a large front-end loader and hauled by trucks to the dump.

Ventilating air is blown into the mines by fans of tremendous capacity. In mines that are hot the air is refrigerated and then circulated through the mine.

At some mines the drilling and blasting cycle has been eliminated by using boring machines for making tunnel and vertical or inclined shafts between levels. Some tunnel boring machines have cutter heads 6.7m in diameter.

Mine roof is supported in many ways. Bolts which penetrate the roof for 1.8 to 4.6m pin rock layers together to form beams to hold loose fragments in place. Gunite, a cement, sand and water mixture, is sprayed on rock walls to inhibit air slacking and subsequent spalling. Chemicals and various cement formulas are pumped into rock to strengthen it by filling fractures and fissures with binding material. Excavations in rock made unstable by ground-water action sometimes are stabilized temporarily by freezing, and this technique also is used at times to control excessive water flows.

In terms of production, the largest hardrock underground mines produce over 45,720 metric tons per day.

In the great open-pit mines, tens of thousands of tonnes of material are moved on a daily basis at many mines. One mine, the Bingham Canyon mine of Kennecott Copper in Utah, excavates 304,800 metric tons per day of ore and waste. The shovels in the big copper and iron ore mines now are as large as 12.2m^3 in capacity. The largest trucks in 1970 had reached a capacity of 203 metric tons. Some are propelled by electric-motor-driven wheels.

Development plans for mining include turbines for powering haulage units, atomic blasting, chemical mining and nonexplosive fragmentation of rock.

The trend toward greater mechanization and higher capacity units has multiplied the efficiency of mining labor manifold, and has been responsible for gains in productive capacity to parallel increased demand for minerals. Some data on productivity are supplied in Table 1-1.

Table 1-1 Various example of productivity

Ancient	1. Egypt: 1900 B.C., 12th Dynasty (from a papyrus). Granite quarried for ten statues 2.64m high by 152mm thick. Task force included 20 soldiers, 30 sailors, 30 specialists and 2,000 laborers. 2. Eighty pyramids required an estimated 358.9 million metric tons of rock. 3. 1370 B.C.: An Egyptian smelter employing 700 men produced that year 250kg of Au and 1,620kg of Ag ingots. 4. 600—500 B.C.: The advance of underground galleries was 12.7m in 24 hr in soft rock. In hard rock, 7.6—9m was advanced per year

Continued 1-1

20th Century

5. In 1968, boring machines made 0.6 to 9meter per hour of tunnel, average 2.4 to 3.7meter per hour. Machine cutterheads were 0.9 to 11m in dia., averaging 3.4m. Maximum daily performance over 122m, 30 to 61m per day not unusual.

6. High-speed shaft sinking in South Africa: 1951 record, 154m per month; 1961 record (still standing in 1968), 384m per month.

7. Raise boring average advance, reported in 1968, 1.5m/h.

8. Mining potash by continuous miners: 711 to 1,016t/h, 1971 report.

9. Cut-and-fill mining with trackless equipment extraction average 51t per manshift, 1971 report.

10. Continuous mining machine rated capacity in coal, 406—508t/h, 1971 report.

11. Offshore tin dredge capacity, 382,275m^3 per month, 1971 report.

12. Average coal output in the U.S. in 1968 from all types of coal and lignite mines was 17.57 metric tons per man-day.

13. Average iron ore output in the U.S. in 1969 from all types of mines was 39.17 metric tons per man-day.

14. U.S. pig iron production, Millions of metric tons

Year	1900	1910	1920	1930	1940	1950	1960	1970
U.S. pig iron production, Millions of metric tons	15.6	31.1	42.1	36.2	48.2	67.5	69.7	93.3

15. World coal production, Millions of metric tons

Year	1900	1910	1920	1930	1940	1950	1960	1970
World coal production, Millions of metric tons	777	1,166	1,320	1,414	1,799	1,805	2,630	3,250

1.1.5 Man, Technology and Mineral Economics

Mineral deposits are not evenly distributed and few nations in the world, if any, are self-sufficient in all mineral commodities. Certainly, the United States seeks abroad many commodities. Iron, tin, chromium, manganese, diamonds and gold are just a few. Table 1-2 illustrates the interdependency of nations in mineral commodities.

Table 1-2 Typical world mineral trade statistics illustrate interdependence of nations

Mineral or metal	World output, 10^3 mt	World exports		Principal exporters	Principal importers
		10^3 mt	Million $, F.O.B.		
Natural gas	780,000	20,000	200	Canada, Algeria, USSR	USA, France, Poland
Crude oil	1,900,000	900,000	10,000	Libya, Venezuela, UAR (Egypt), Iran, Kuwait, Iraq, USSR	Japan, Italy, FR Ger, UK, France, USA
Refined oil	1,800,000	320,000	8,000	Venezuela, Antilles, Italy, Trinidad, Netherlands, UAR (Egypt), UK, USA	USA, UK, FR Ger, Sweden, Japan
Coal	2,780,000	150,000	2,000	USA, Poland, USSR, FR Ger, Australia	Japan, France, Italy, FR Ger, Belgium
Iron ore	680,000	235,000	1,500	Canada, Australia, USSR, Sweden, France, India, Venezuela	Japan, USA, FR Ger, Belgium, UK, Czechoslovakia

Continued 1-2

Mineral or metal	World output, 10^3 mt	World exports		Principal exporters	Principal importers
		10^3 mt	Million $, F. O. B.		
Pig iron	380,000	9,500	400	USSR, India, FR Ger, Finland, S. Africa, Canada, Poland	Japan, E Ger, Poland
Scrap iron	…	15,000	600	USA, France, FR Ger, UK	Italy, Japan, FR Ger, Canada, Mexico
Steel products	400,000	62,000	7,500	Japan, FR Ger, Belgium, USSR, France, UK	USA, FR Ger, France
Aluminum	8,400	1,900	1,000	Canada, Norway, USSR, France, FR Ger, Hungary	USA, UK, FR Ger, Japan, Belgium, France, Italy
Alumina	20,000	4,700	1,000	Jamaica, Hungary, Guyana, France	USA, Norway, Canada, Ghana, Poland, Japan, Spain
Bauxite	45,000	20,500	200	Jamaica, Surinam, Australia, Guyana, France, Yugoslavia	USA, Japan, Canada, FR Ger, USSR, Italy, UK, France
Refined copper	6,650	2,270	2,000	Zambia, Chile, Belgium, Canada, USA	UK, USA, FR Ger, France, Italy, Japan
Blister copper	5,880	820	700	Chile, Congo, Peru	USA, Belgium, FR Ger, UK
Copper-in-ore	5,460	440	300	Canada, Philippines, USA, Chile	Japan, FR Ger, USA, Spain
Refined lead	3,530	935	300	Australia, Canada, Mexico, USSR, Peru	UK, Italy, France, FR Ger, Netherlands
Lead-in-ore	3,060	600	200	Canada, Peru, Australia	Belgium, FR Ger, Japan, France
Refined zinc	4,710	1,250	400	Canada, Belgium, Australia, Japan, USSR, Poland	USA, FR Ger, UK, India
Zinc-in-ore	5,070	1,970	600	Canada, Peru, Australia	Belgium, Japan, France, UK
Manganese ore	18,000	8,500	300	S Africa, Gabon, USSR, Brazil, India	France, FR Ger, Norway, UK
Nickel	500	350	800	Canada, N Caledonia, UK, Norway	USA, Japan, FR Ger, UK, France
Tin	230	170	500	Malaysia, Bolivia, Thailand, Indonesia	USA, Japan, UK, FR Ger, France
Tungsten-in-ore	32	13	70	S Korea, Bolivia, China, Australia, Portugal	UK, FR Ger, Japan, Poland, Australia, Sweden
Cement	510,000	13,000	200	USSR, Japan, Belgium, FR Ger, Taiwan, France	Many countries
Fluorspar	3,500	2,000	60	Mexico, China, Thailand, Spain, France, Italy	USA, Japan, FR Ger, Canada, USSR
Phosphate rock	84,000	35,000	300	USA, Morocco, USSR (apatite), Tunisia, Nauru, Jordan, Christmas	Japan, Australia, France, FR Ger, Italy, Canada, UK

Continued 1-2

Mineral or metal	World output, 10^3 mt	World exports		Principal exporters	Principal importers
		10^3 mt	Million $, F. O. B.		
Potash (K_2O)	15,500	6,500	300	Canada, E Ger, USSR, France, USA, FR Ger, Spain	USA, Poland, Japan, UK, Belgium
Salt	125,000	12,000	100	Mexico, Netherlands, FR Ger, China	Japan, USA, Belgium
Sulfur	19,000	6,500	300	Canada, USA, Mexico, France	USA, UK, Australia, India, France, Netherlands, Belgium, FR Ger, Hungary

Sources: World Bureau of Metal Statistics, World Metal Statistics (monthly), 1970, London; Institute of Geological Sciences, Statistical Summary of the Mineral Industry, 1963—1968, London, 1970; U. S. Bureau of Mines, Minerals Yearbook, 1967—1968.

Compiled by K. P. Wang, U. S. Bureau of Mines, and presented in his report "Minerals and Metals in International Trade" to The United Nations Minerals Economics Seminar, Ankara, Turkey, 1970.

Note: Mt stands for metric ton(s). F. O. B. gives free on board.

Individual mineral deposits become valuable for several reasons. They may be very rich, they may be more accessible than other deposits, the location may be more favorable, or the commodity is in great demand. A few examples may be necessary to explain that the value of a deposit is not so much intrinsic as is the relevance placed upon it by man.

The Cerro Bolivar iron ore deposit in Venezuela was noted in the 1920s but it was not of importance until the available quantities of near-surface iron ores of Birmingham began to diminish. It then was found that Venezuelan ore, being high grade and on the surface, could be brought to Birmingham more cheaply than the local ores.

The bastnaesite deposit (a mineral valuable for its content of rare earths) discovered at Mountain Pass, Calif., in the early 1950s did not assume any great importance until color television was in vogue in the 1960s because this created a demand for the europium contained in bastnaesite.

A sand and gravel deposit 161m from any habitation may not have any economic value because of deposits which are closer to a settlement. However, it is necessary to go farther and farther away from cities to find these construction ingredients as nearer deposits are exhausted or environmental considerations prohibit mining.

Anthracite in northeastern Pennsylvania at one time was of great economic importance for the home-heating market. Today, much less anthracite is mined because people prefer oil or gas for domestic heating and no large new use has been found for it.

Obviously, technologic advances have made deposits of lesser mineral content into economic deposits. In 1900, copper ores in the U. S. were not mined unless they contained 6% copper. Today the average grade of ore mined is about 0.70% copper.

Uranium was not sought prior to 1942. Today, it is the product of many mines in New Mexico, Utah, Wyoming, Colorado and other states and countries.

The United States has large reserves of manganese that are not economic at this time because foreign ores are available at lower cost. Someday they will yield a valuable commodity.

The oil shale of the Colorado Plateau will be mined and treated when cheaper petroleum becomes scarce.

In two outstanding examples, that of nickel and molybdenum, the mining company created a demand by developing uses for a commodity that was not previously needed.

In the field of industrial minerals, the market development approach to transforming a useless mineral curiosity into an economic deposit is not uncommon.

Advances made by the mining engineer in technology are an important factor in determining the demand for mineral commodities. However, the chief role of the mining engineer is to extract the minerals presently needed in the most efficient way for the greater well-being of society.

1.1.6 Conservation in Mining

Mineral deposits are a wasting asset. The minerals are irreplaceable, so that as the deposit is mined it decreases in value. The miner, unlike the manufacturer or the farmer, must discover new sources of raw materials to remain in business.

The prudent miner conserves his ore by planning mining in such a way as to maximize his profit from it over the life of the deposit. In mining, it is essential to view the financial outcome on the basis of the full life of the operation as nearly as this can be anticipated, which is not necessarily the case in other businesses. The economics of mining are influenced by efficient extraction. This usually means taking out ore of lower metal content to the greatest extent possible along with better grade ore.

Mineral conservation is achieved through several different approaches. The foregoing refers to the economic approach, which is an important technique. A second technique is competent geologic analysis to map the deposit and locate new ore. This is done so successfully that many mines are able to replace the ore mined year after year by the discovery of new reserves.

Technical research has played a tremendous role for generations in conserving minerals. New methods and new machines make it possible to extract ores of lower and lower grade. It is well known that the waste piles of old mines and metallurgical facilities are reworked to remove metals that formerly could not be recovered. Mining techniques, such as improved roof support, permit the reduction in size of pillars composed of ore so that more complete extraction is possible. The use of leach solutions at mines permits recovering metal from material which is too low grade to process in any other way. Metal is also extracted from natural mine drainage water.

Technology has increased conservation of minerals by enabling such minute metal values as molybdenum and rhenium to be recovered from copper ores. There are many other examples of by-products, scarce in volume, being taken from ores.

The provident miner attempts to utilize as much of the material he extracts as he can. Rock refuse may be prepared for construction use, and mill sand often is returned to an underground mine to prevent caving. Even the opening may be used subsequently for storage of gas or other materials.

Less spectacular as an aid to conservation but of extreme importance is the day-to-day good operation of a mine as a result of experienced supervision, a stable work force and the avoidance of work interruptions.

1.1.7 Influence of Mining on Environment

It is not usually possible to extract minerals from the earth without changing the natural environment in some way. A mine requires access roads, power and water. An opening in the ground must be made. Usually, in the case of underground mines, the surface disturbance is small compared to open-pit mines. Mine site acreage must be devoted to processing facilities, shops, offices, change house and storage facilities. The waste materials from processing operations must be disposed of. These may be solid, liquid or gaseous. In addition, there is the atmosphere in the mine and other facilities that must be controlled to safeguard the workers' health.

Environmental controls have been applied to operations for decades. These include land restoration, water purification, dust suppression and diffusion of noxious gases. The techniques for these controls have been developed over a long period of time and, as the technology improves, the adverse effects of mining on the environment will continue to be reduced. At the present time, environmental legislation being proposed and enacted poses a problem to mine operators of a greater or lesser magnitude depending on conditions at each particular property. These problems are the availability of methods and machines to achieve the desired pollution control, the time factor to meet deadlines and the effect on the operation of the additional cost.

Since the demand for minerals and fuels increases from year to year partly because of population but also because of rising living standards in developed and underdeveloped areas, it will be necessary for society to continue to cope with the extractive industries. Paradoxically, products of the mineral industry are required in environmental control machinery and processes.

In the controversy over utilization of land for mining vs. other applications, there is the plausible solution of orderly multiple use. Where this concept is in practice, minerals are extracted in a way to minimize environmental nuisance, followed by restoration for other uses of the land. Examples are in coal stripping and phosphate mining where the land is restored in preparation variously for urban use, recreation, timber or grazing.

1.1.8 Future of Mining

Dependence of man on his mineral heritage, long established, will continue indefinitely. Requirements of technologic progress will place increasing demands upon the mineral industry to produce more and a greater variety of elements. This demand will be met by greatly improved exploration techniques based on better understanding of the genesis and localization of mineral deposits and methods of locating the presence of mineral concentrations deep within the earth. Progress in technology will make ore out of many rocks which are considered useless today.

Methods will change in a way to make mining safer as well as reduce environmental impairments.

The technology of the exploration geologist, the mining engineer and the metallurgist is benefiting the work of scientists and engineers in other disciplines. For example, great benefits to other groups can be derived by study of accomplishments in environment made by mining engineers, one of which was the development of air conditioning.

The mining community at the same time is benefiting from knowledge developed in other fields. This cross pollination is accelerating progress.

Research is going on in government, private research foundations, universities and in industry. A substantial amount of this work is mineral-oriented.

The arts and sciences of the miner began with early man. They are changing rapidly in this period but assuredly mining will continue as long as Man does on earth.

1.2 Non-ferrous Metal Deposits

1.2.1 Non-ferrous Metals

There exist several classifications of chemical elements which differ in defining the group of metals assigned to the non-ferrous category. In any case manganese and chromium, besides iron, are usually assigned to the ferrous metals group. Sodium and potassium, which are encountered in nature as chloride salts, as well as calcium, strontium and barium tending to nonmetals are commonly excluded from the consideration. We will adhere to the classifications which take into account the geological, mineralogical and processing properties of metals (Table 1-3).

Table 1-3 Geological, mineralogical and technological classification of non-ferrous metals

Class	Periodic table group							
	I	II	III	IV	V	VI	VII	VIII
Heavy	Cu	Zn, Cd		Pb				Ni, Co
Minor		Hg		Sn, Ti, Zr	As, Sb, Bi			
Light	Li	Mg, Be	Al					
Refractory					Nb, Ta, V	Mo, W		
Noble	Ag, Au							Pt, Os, Ir, Ru, Rh, Pd
Rare earth			RE, Sc, Y					
Dispersed	Rb, Cs		Ga, In, Tl	Ge, Hf				
Radioactive		Ra	U, Th, Pu					

The non-ferrous metals industry was among the most high-level sectors of the FSU (Former Soviet Union) heavy industry engaged in the mining of ore deposits, including ore mining, concentration and metallurgy with production of non-ferrous metals for further processing. After World War II a sector whose challenge included mining and processing of ores of uranium and other radioactive elements, and later on beryllium and lithium, was separated with a view to creating an atomic bomb.

More than three hundred mines, concentrators, mining and metallurgical complexes, and metallurgical works of various capacity have been in operation in the above sectors; each of them employed from several hundreds to tens of thousands of workers and specialists.

The main products were concentrates, cast metals, alloys, chemical compounds, non-ferrous metals, and alloys products. By-products (sulfuric acid, building materials, some fertilizers and microfertilizers) used to be made from the waste in many cases. The entry raw materials were for the most part comprehensive, and associated components were extracted from ore on numerous occasions. For example, in the lead and zinc sector additional 18 components alongside lead and zinc used to be extracted, with more than 40 saleable product types made on their base. Fifteen components were recovered from copper ore with 40 product types made. About 30 per cent of sulfuric acid in the country were produced in the non-ferrous metals industry.

By the end of the 80s the run-of-mine material amounted to 2.5 billion tonnes, with virtually the same waste tonnage. The basic ore dressing process (about 90%) was flotation. Metallurgical operations were at a fairly up-to-date level. By contrast, environment control at the plants and works of the sector has led to an ecological catastrophe.

In order to meet the requirements of the national economy and the military industry, the Former USSR Ministry of Non-ferrous Metals Industry used to carry out measures on the development of the non-ferrous metals industry and the production build-up.

The Ministry's terms of reference were:

(1) Start-up and commissioning of operations and facilities;

(2) Renovation and modernization of plants and works;

(3) Final exploration of deposits, mining operations in open pits and mines, shaft sinking;

(4) Preparation of standards on estimating explored mineral product reserves, registering state and movement of reserves and depletion of mineral resources in the Earth's interior;

(5) Comprehensive ore processing with output of non-ferrous, rare and precious metals as well as diamonds;

(6) General contractor's functions with respect to non-ferrous metals plants under construction abroad with Former USSR technical assistance.

1.2.2 Deposits and Ores of Non-ferrous Metals

The virtually unlimited reserves of mineral resources in the territory which occupied one sixth of the Earth's land were the raw materials base of the Former USSR non-ferrous metals industry. Essentially any genetic deposit type is encountered here. We shall discuss two radically different deposit types: geologic and technologic.

A mineral resource deposit on the Earth's surface or in the interior which can be used in industry in quantity, quality, occurrence conditions and processability is called a geologic deposit. Deposits whose usage is economically feasible at the existing technology and engineering progress level are assigned to the workable category.

Geologic deposits are composed of one or several orebodies whose generation takes place throughout the whole process of geological evolution and alteration of the Earth's crust in several stages. The concentration of ore-forming minerals of a certain composition under certain geological and physicochemical conditions prevailing at some time period, is an ore-forming stage. The number of these stages usually amounts to four or six, sometimes ten during the orebody generation process.

The ore-forming stage accounts for the whole mineral accumulation period which combines a number of successive phases belonging to one and the same genetic process. The ore deposit is usually composed within one mineral accumulation stage, more seldom two or more. For example, the upper parts of the ore bed may contain the mineral mass of the primary hydrothermal stage and of the stage predetermined by the secondary oxidation of the ore near the Earth's surface. The ore bed may be also formed in several stages of a monotype process common to different geological epochs.

Regional regularities and geological factors of mineral resource deposit formation were the result of endogenic and exogenic processes during the evolution of the Earth's crust. Substances required for mineral resources to be formed enter from the upper mantle of the crust and from the surface in magmatic melts, liquid and gaseous solutions. Magmatogene (endogenetic) deposits are subdivided into several groups.

Magmatic deposits are formed during impregnation of magmatic melts into the crust and cooling-down of them. The ores of Cr, Fe, Ti, Ni, Cu, Co, the platinum metals group, etc. are associated with the intrusions of the basic composition; the ores of Pd, Ta, Nb, Zr and rare earths (RE) are confined to alkaline massifs of magmatic rocks. The deposits of mica, feldspar, precious stones, Be, Li, Cs, Nb, Ta, partly Sn, U and RE are connected with granite pegmatites by genesis. Carbonatites associated with ultrabasic-alkaline rocks form Fe, Cu, RE, apatite and mica deposits. There are ores of Fe, Cu, Co, Pb, Zn, W, Mo, Sn, Be, U, Au, Bi, rock crystal, graphite and other mineral resources in contact-metasomatic deposits, especially, in scams. Cu, Ni, Co, Zn, Pb, Bi, Mo, W, Sn, Li, Be, Ta, Nb, As, Sb, Hg, Cd, Ln, S, Se, Au, U, Ra, as well as quartz, barite, fluorite, asbestos, etc. are contained in pneumatolytic and hydrothermal deposits.

Sedimentary deposits which are formed in exogenic processes are subdivided into sedimentary, placer and weathering. Sedimentary deposits are generated on sea, lake, river and marsh bottoms with formation of stratified beds in enclosing sedimentary rocks. Placers containing Au, Pt, diamonds, etc. are accumulated in coastal sediments of seas and oceans as well as in river or lake sediments and on valley slopes. U, Cu, native sulfur ores and sedimentary deposits of Ni, Fe, Mn, bauxite, magnesite, kaoline are connected with the ancient and modern weathering crust of which infiltration deposits are typical. Previously available deposits undergo transformation deep in the interior under the conditions of high pressure and temperature, with the appearance of metamorphogenic beds and marble, andalusite, cyanite, graphite and other deposits newly formed during the rock metamorphism process.

The industrial application field is used to distribute mineral resources into:

(1) Fuel and energy resources (oil, natural gas, fossil coal, combustible shale, peat, uranium ore);

(2) Ore resources which are the raw materials base of iron and steel and non-ferrous metals industry (iron and manganese ore, chromite, bauxite, polymetallic ore, copper-nickel, tungsten, molybdenum, tin, precious metals ores, etc.);

(3) Mining and chemical raw materials (phosphorite, apatite, table, potassium and magnesia salts, sulfur and compounds, barite, boron ore, bromium and iodine-containing solutions);

(4) Natural building materials and non-metallic stones (marble, granite, jasper, agate, rock crystal, garnet, corundum, diamond, etc.);

(5) Hydromineral resources (underground sweet and mineralized water).

The grouping of mineral resources is a matter of convention, since the industrial application fields for one and the same mineral resource may differ and vary with technological advance.

Certain formations of rocks and mineral resources associated with them occurred at successive stages of the Earth' s crust evolution. Their repeatability in the crust evolution history brought about repeatability in formation of similar groups of mineral resources from the most ancient to the jungle stages of geological history marked by metallogenic (or mineragenic) epochs.

Non-ferrous heavy metals ore fields are represented by copper, lead, zinc, nickel, antimony ore deposits. Metal reserves in the largest of them amount to tens or hundreds of millions of tonnes with the usual metal content of per cent units in the deposits. A large part of copper ore is available from stratiform cupreous sandstone and shale deposits. A large source is also the hydrothermal stockworks of the so-called copper-porphyry ores. Copper ores are also mined from volcanogenic pyrite and hydrothermal vein deposits. A substantial part of this metal output used to be extracted from magmatic sulfide copper-nickel fields.

In nature lead and zinc are usually encountered jointly in the composition of polymetallic ore deposits. Stratiform sheet-like deposits in carbonate rocks play a large role among them. Besides, lead-zinc ores are mined from volcanogenic pyrite deposits of hydrothermal metasomatic fields in carbonate rocks both among scams and without them, hydrothermal vein deposits. The major part of nickel ore is mined from magmatic sulfide copper-nickel deposits and from silicate composition weathering deposits. All antimony ore deposits are attributed to hydrothermal stratified and vein ones.

The biggest reserves of tin, tungsten, molybdenum, mercury, beryllium, lithium, tantalum and niobium amount to hundreds of thousands of tonnes with metal content in ore of usually not more than 1 per cent. A large proportion of tin ore is mined during working of hydrothermal sulfide-cassiterite and quartz-cassiterite deposits. Tungsten ore is contained in hydrothermal vein and stockwork wolframite deposits as well as in scams deposits.

Noble metal ore deposits are represented by gold, platinoid and silver categories. Their reserves commonly amount to tens or hundreds (seldom thousands) of tonnes with the content of, for example, gold which seldom exceeds 10 g/t, i. e., 0.001%. The most abundant gold-ore deposit type is gold-bearing quartz and other composition hydrothermal veins and stockworks. Substantial value is attached to mining of gold from volcanogenic hydrothermal complex gold-silver ores.

Uranium, thorium and radium ore deposits belong to radioactive ore deposits. In some deposits the uranium ore reserves amount to thousands or tens of thousands of tonnes, seldom more, with a conventional metal content of tenth parts of per cent in ore. The role of sedimentary deposits usually complicated by near-surface infiltration processes is rather large among uranium deposits. Hydrothermal metasomatic and vein uranium ore deposits are various.

Dispersed element ore deposits (actinium, hafnium, germanium, indium, cadmium, protactinium, rhenium, rubidium, selenium, scandium, thallium, tellurium, caesium, etc.) are represented by sedimentary, magmatic and metamorphic series deposits. Dispersed elements are extracted as by-products during processing of other metal ores.

There are no independent deposits of rare-earth elements of the cerium and yttrium groups, ores of these elements are also for the most part extracted as by-products during mining of magmatic, carbonatite, albite, hydrothermal and placer deposits of non-ferrous, rare and radioactive metal ores.

1.2.3 Technologic Deposits

A technologic deposit is an accumulation of mineral substances on the Earth's surface or in workings which was formed owing to separation of substances from solid massif and storage as waste of mining, dressing, metallurgical and other operations and which is suitable for commercial use in quantity and quality (for extraction of metals and other valuable components, generation of fuel and manufacture of building materials).

The deposits include mineral mining dumps, concentrator tailings dumps, ash and slag disposal areas of power-and-heat plants, stored waste of metallurgical and other operations. The so-called industrial waste burial polygons, which serve as dumping grounds for used electric and electronic equipment, automobiles, machinery, devices, and finally municipal waste ("urban res") are classified as technologic deposits and provide a universal source of many rare and dispersed elements. For example, the main source of germanium is power-and-heat plant ash, of rhenium -roasted molybdenum concentrate dust, of selenium and tellurium -polymetallic ore, of gallium-bauxite and nepheline processing waste. Technologic deposits become more and more important as a source of many mineral types. In industrialized developing countries up to 80% on the average of non-metallic building materials are produced from overburden, up to 20% of copper (more than 30% in the USA) from oxidized copper ore dumps and concentration tailings by bacterial and acid leaching. The total content of valuable components accumulated over 20—30 years in technologic deposits is comparable to and sometimes exceed their quantity in annual routine ore output.

Just like conventional deposits, they have a certain valuable component distribution structure, zones of secondary hypergenesis, oxidation, segregation, etc., but unlike natural (geologic) deposits-a lower valuable component content. Waste is dumped so as to provide more favourable conditions for their downstream recycling (stable slopes, minimum dumping costs, flowline process application). Waste utilization is as a rule economically feasible, since the deposit is on the surface and the material in it is partially disintegrated. However, these deposits must be utilized at the proper time, because in many cases they undergo substantial modifications, losing processability and practical value. For example, deposits containing heavy non-ferrous metal sulfides may be depleted up to total loss of the metals owing to spontaneous bacterial leaching processes. Separate storage is required in some cases for value improvement during dumping, otherwise a technologic deposit may not form. Separate dumping entails certain expenses and noticeably increases mining costs in some cases.

1.2.4 Extraction Methods of Valuable Components

Deposit mining results in output of ore, a natural mineral product containing metals or their compounds in quantity and form suitable for commercial use. Even naturally high-grade ore requires dressing as a rule. Non-ferrous metal ores are polymictic and accompanied by commercially useless minerals. Ores are considered complex if they have associated valuable components (metals, non-metals), whose recovery is economically feasible. Judging from the chemical composition of prevalent minerals, ores are divided in silicate, siliceous, oxide, sulfide, carbonate and mixed. All ores mined from deposits enclosed into native sedimentary, magmatic and metamorphic rocks are called native, while ores produced from river, lake, sea and ocean sands are called placer.

Ore structure is dictated by combination of individual mineral grains in an ore mineral aggregate. Structures are differentiated into equigranular, inequigranular, oolite (with concentrically rounded mineral accumulations), porphyritic (with individual coarse mineral grains in an equigranular mass), radiated etc. Ores are divided into those with uniform, nonuniform and extremely nonuniform composition according to the valuable mineral distribution nature.

For ore mining and processing of essential importance are physical properties: hardness, strength, jointing, density, melting point, magnetic and electromagnetic properties, conduction, radioactivity, solubility, sorption, and granulometric composition.

Valuable ore components are the constituent of the mineral resource, whose extraction is technically and economically feasible with the view of commercial application.

Essential and associated (accessory) valuable components are distinguished. Essential valuable components are contained in mineral resources in commercial concentrations, which determine the basic value, application and name of the resource. When two or several essential components are available, the mineral resource is characterized as a complex one (for example, copper-molybdenum or copper-lead-zinc ore). Associated minerals and dispersed elements are mineral constituents, whose extraction is economically feasible only jointly with the essential valuable component. Valuable minerals are incorporated into the mineral resource composition in low concentrations and can be extracted to selective concentrates or accumulated in essential valuable component concentration products. Dispersed elements are contained in the composition of essential or associated valuable components as isomorphic, finely-dispersed or other impurities (usually in low concentrations) and can be extracted only during metallurgical or chemical concentrate separation.

Complete comprehensive recovery and utilization of all valuable components is a condition for efficient use of the earth's interior. The notion of valuable component is not rigid, but varies with industrial requirements, progress in engineering and technology of mining, dressing and processing of mineral products.

The commercial content of an essential component in a mineral resource at which extraction and use are economically feasible is of substantial significance. The commercial content for any mineral product type may vary over a wide range depending on economic-geographic deposit conditions, geological structure, ore composition and properties, mining and processing technology and equipment, ecological requirements. Mineral product standards establish limiting values of the commercial content, on whose basis mineral reserves are estimated.

Combined processing of mineral products, which is a production process composed of different dressing and metallurgical techniques and operations, affords the most efficient selection of components. Selection of components by properties (difference in density, magnetic susceptibility, conduction, adsorption, wettability, etc.) is based on creation of a gradient of mineral particle, ion or molecule concentration in a liquid or gas phase as well as at interfaces using different fields of forces and actions: magnetic, electric, gravity, adsorption, etc. Combined processing is carried out by several properties through combination of fields applied in a single apparatus (combined process) or in a line of sequentially arranged apparatus (combined flowsheet).

Combined dressing of mineral products takes place without alteration of the phase and chemical composition of the minerals contained in the ore, i. e., it results from a combination of dressing methods. Combined flowsheets are conventionally composed of primary gravity operation (heavy-media separation, jigging, screw separator dressing, sluicing, etc.) and downstream magnetic separation or flotation. These flowsheets are typical of iron (gravity-magnetic separation), manganese (gravity-flotation) and rare-metal ores (gravity-magnetic or electric separation-flotation). Widely-used combined dressing processes include gravity flotation: concentration table flotation (diamond ore), flotation jigging (rare-metal ore). There are also magnetic field classification (magnetite, titanomagnetite ores), flotation in a magnetic field, run-current field, vibration or ultrasound field.

1.2.5 Deposit Evaluation and Mineral Resources

Mineral resources are a totality of useful materials in which chemical elements and minerals formed by them are available at much higher concentration compared to their abundance ratio in the Earth's crust. They are nonrenewable natural resources. The part of mineral resources prepared for mining is called a mineral product base, quantitatively evaluated by geological prospecting and exploration. Preparation of mineral resources for mining includes formation of a mineral product base of mining industry by means of geological exploration in order to evaluate commercial reserves of mineral resources at available natural sites and registration of technologic deposits as mineral sources as well as setting of mining and processing undertakings in these regions.

Prospecting and exploration costs are commonly 1.5%—5.0% of the cost of the metal produced (iron and steel and non-ferrous metals), more seldom 20%—30% (some non-ferrous and rare metals), mining and dressing costs range from 6—10 (for iron or aluminum) to 70%—80% (for some non-ferrous and alloying metals) of the total product cost.

Mineral production is in continuous growth, and an especially large scale was reached in the 20th century. The 1901—1980 period accounted for 90%—95% of the total mining output of tin, lead, zinc, silver, mercury and antimony ores, as well as diamonds, 87% of iron ore, 85% of copper and 70% of gold ore. At the same time the number of mineral resource types increased drastically, and commercial production of bauxite, titanium, molybdenum, tantalum, niobium and other ores began.

Mining output grew irregularly. The periods from the end of the 19th to the beginning of the 20th centuries and from the middle of the 50s to the middle of the 70s stand out as the time of high rates of output increase in mining of many mineral resource types. For example, in the 1950—1975 period production of niobium concentrates increased more than 20 times, bauxites 11 times, nickel and titanium concentrates 5—5.5 times.

For many types of mineral products output growth was mostly due to working of large and very large deposits. Higher mining rates involve working of deposits with leaner ores and of new geological-industrial types. For example, at the end of the 19th century the most of the run-of-mine copper ore contained 5%—8% of copper, but this figure fell to 0.7%—3% in 1925 and to 0.4%—1.5% in the late 70s. Active mining of hitherto unworked large copper-porphyry type deposits began in the 50s.

Increased mining output brought about depletion of near-surface fields and larger mining depths which increased from several tens and hundreds of metres to 1,000m on the average for non-ferrous metal ores (2—3.9km at particular undertakings). Fields located in remote regions, for example, at Arctic latitudes and on the sea shelf, are being drawn into mining. Increased complexity of the geological conditions, higher power and investment requirements lead to objective increase in the costs of exploration, mining and primary processing of the mineral resources. However, the trend towards higher per-unit cost is largely offset by larger production output and the relative increase in the cost of other goods. Thus, the per-unit costs expressed in terms of prices for other products and services do not exceed those calculated in a similar way for the 1980s when the copper content was 5%—8%. An objective aspect is the increase in environment control costs (8%—20% of all investments in mining in industrialized countries).

Seventy-eighty per cent of the reserves for each mineral resource type in the mineral product balance account for a comparatively small number of large deposits and gigantic fields, while the other part is available in medium and numerous small deposits. Commercial value and reserve amount serve conventionally for distinguishing unique deposits of major importance in the world reserves as a whole, large deposits in large countries rich in mineral resources, medium deposits in medium and small regions of large countries, small and minute deposits in individual districts and undertakings. Table 1-4 presents the reserves of individual types of mineral resources by continent.

Table 1-4 Allocation of main mineral resource① types by continent and country group

Mineral resource		Europe	Asia	Africa	North America	South and Central America	Australia and Oceania	Total	Developing countries	Industrialized countries
Bauxites	bln t	0.25	1.83	12.35	0.04	3.9	2.42	20.8	18.2	2.6
	%	1.2	8.8	59.4	0.2	18.8	11.6	100.0	87.3	12.7

Continued 1-4

Mineral resource		Europe	Asia	Africa	North America	South and Central America	Australia and Oceania	Total	Developing countries	Industrialized countries
W ore②	mln t	159	267	32	367	116	119	1,060	394	666
	%	15.0	25.2	3.0	34.6	11.0	11.2	100	37.2	62.8
Co ore③	mln t	0.05	0.43	1.8	0.26	0.1	0.3	2.94	2.51	0.43
	%	1.7	14.6	61.2	8.0	3.4	10.2	100	85.3	14.7
Cu ore	mln t	8.3	47.6	78.9	122.5	174.3	15.8	447.4	302.8	144.6
	%	1.9	10.6	17.6	27.4	39.0	3.5	100	67.7	32.3
Mo ore②	mln t	—	0.2	—	4.2	2.85	0.15	7.4	3.2	4.2
	%	—	2.7	—	56.8	38.4	2.1	100	43.1	56.9
Ni ore③	mln t	2.1	11.9	5.2	10.1	4.9	16.0	50.2	32.1	18.1
	%	4.2	23.8	10.4	20.0	9.8	31.8	100	64.0	36.0
Sn ore③	mln t	0.2	1.8	0.4	0.1	0.8	0.2	3.5	2.9	0.6
	%	5.7	51.5	11.4	2.9	22.8	5.7	100	82.9	17.1
Pb ore③	mln t	12.9	8.3	11.1	47.0	13.8	23.0	115.6	25.5	90.1
	%	11.1	7.2	9.6	40.7	11.5	19.9	100	22.1	77.9
Zn ore②	mln t	23.3	28.1	24.8	60	21.4	21.0	178.6	56.7	121.9
	%	13.0	15.7	13.9	33.7	12.0	11.7	100	31.7	68.3

Note: Bln t means billion t and mln t, million t.

①Identified and proven reserves as of early 1984 (industrialized and developing countries).

②In terms of oxide.

③In terms of metal.

Despite the depletion of individual long-exploited deposits and reduction in explored reserves of mineral resources in some countries, mining levels reached in the early 80s are secure for the visible future. However, a large part of the identified mineral resources is concentrated in deposits of relatively lean or deep ores and under complex geological conditions.

Commercial development of mineral resources includes their evaluation (geological prospecting and exploration) and working (mining, dressing and processing), whose scope and intensity are dictated by the peculiarities of industrial and socio-economic development, together with position of the mineral product sector in the national economy. Mineral resources are nonrenewable, which dictates judicious utilization and reduction of losses during mining, processing and handling as well as utilization of the secondary materials and compliance with an ecological and economic approach in exploitation of the mineral resources.

Mineral exploration is carried out with a view to estimating commercial value of mineral deposits given a positive rating in a prospecting and evaluation process. A reserve of replenishing exploration projects are also deposits evaluated in the past, but assigned outside the balance for variance reasons. Their re-evaluation on the basis of new geological-genetic concepts, situation changes, more perfect exploration techniques along with innovative technologies of mineral mining and processing, permits transfer of some of the projects for preliminary exploration without additional field work.

Exploration establishes the geological and industrial deposit parameters required for commercial valuation, planning of a mining enterprise, initiation of the mining and processing of the mineral resources to be extracted. For example, the morphology of mineral bodies is determined in this way, which is of primary importance in selection of a future mining system. The contours of the resource body are established with allowance to the geological boundaries, and average content of essential and associated components, availability of harmful impurities, nature of mineral resource distribution, etc. are set in accordance with the data of deposit testing. The data on valuable component content are the initial basis for substantiation of estimation standards and ultimately for solving the problem of commercial value of the project under exploration. These data are equally useful in determining the processability of the mineral resources. Ore deposits are studied stage-by-stage, with more and more detailed exploration through regular densification of the prospecting hole network pattern, with more and more accurate estimation of the reserves and more and more detailed and complete technical and economic evaluation of deposits under study.

Deposit exploration methods depend on a set of relevant engineering facilities which provide maximally complete information. The key operation in prospecting is drilling: percussion (in placer survey), column (core or coreless) and deep; deep prospect holes, shallow shafts, aditus are applied in extra-complex cases (as a rule in exploration of non-ferrous and rare metals ore). The purpose is verification of the prospecting drilling data, clarification of the structure for the most complex deposit areas, technological sampling. Detailed and final exploration is commonly carried out and is grounded on drilling; deep prospecting and prospecting-operating shafts are also sunk at certain projects. The main types of exploration operations are openings (level, vertical and inclined) and underground (usually short) holes: column and perforator (coreless). In order to obtain maximum information on deposit structure and regularities of useful mineral arrangement at minimum cost, the prospecting holes are distributed so that they cross the whole promising zone (horizon, structure) thickness while the prospecting profiles (prospecting intersection groups) are localized primarily across the zone strike. Prospecting network density depends on the degree of the complexity of the geological deposit structure according to which reserves are estimated. As the preliminary survey moves on to exploration and further on to operation the prospecting network becomes denser. In the case of vein rare-metal ore deposits the prospecting network can be condensed to several tens of metres and even more.

Different geophysical and geochemical methods are applied at all prospecting stages. The main purpose of geophysical works is to find out the deposit structure through clarifying the position and following the ore-controlling and ore-localizing surfaces as well as through correlation of prospecting intersections. Geochemical methods are applied for delineation of horizons, zones and blocks (with abnormal distribution of indicator elements) promising in prospecting for buried ore beds.

The choice of engineering facilities for deposit prospecting is dictated by work type: there may be ditchers and shovels for ditch and trench digging, bulldozers for stripping and baring of bedrock, as well as trench exploration of placers, mobile equipment for sinking shafts and mapping-reconnaissance boreholes, descent-ascent mechanisms for sinking deep shafts and holes, drilling machines of different types for prospecting holes, sampling-analyzing devices, computerized geophysical field instrumentation for operational data processing, etc.

Summary horizon-by-horizon plans and sections, which are the basis for estimating mineral reserves, are made up from the primary geological documentation and testing data. The same data are used for composing detailed geological-structural maps, which in turn serve as basis for prognostic maps are constructed. These are the foundation for further projects.

Evaluation of mineral deposits, which is establishment of their commercial value, using combined geological methods (geological evaluation) and economic calculations (economic evaluation) is carried out at all stages of study, prospecting and commercialization, from regional prediction to complete depletion.

The problems of commercial value of the identified deposit, total mineral resource scope and quality (including processability, possible working conditions, etc.) are solved during prospecting. Explored reserves data are used in technical and economic calculations which allow determination of the main indicators of deposit commercialization in several standard versions, taking into account production scope and other parameters——all of which is reflected in a special technical and economic report. Both the geological evaluation of the deposit under study and its region as well as economic assessment are of equal importance at the stage of preliminary prospecting. The former consists in detailed study of mineral localization conditions, impartial estimation of the reliability of the identified reserves and establishment of all potentials for their further expansion; the latter consists in estimation of all cost items (setting-up of mining undertaking, roads, power lines, infrastructure, etc.) which are to be outlaid during commercialization of the project, along with the expected economic efficiency. At the stage of detailed prospecting one main task of the geological evaluation is to determine the degree of reliability of the explored and estimated reserves with a view to exclusion of negative consequences in case of deviation from subsequent mining data.

The morphology of the mineral bodies, their space location, quality and processability, the technical and economic conditions of mining, etc. are finalized. Geological-economic evaluation of deposits under exploration is based on estimation of the identified reserves in turn based on standards. Technical-economic substantiation of estimation standards is based on wholesale prices of the corresponding mineral product type, the shortage level, working conditions, processability and other data.

As the deposit is prepared for commercialization, multi-variant comparative economic evaluation and selection of the most economically profitable project from the explored single-type fields are carried out with a view to determining the profitability level of mining the explored mineral bed, with the time factor included.

1.3 China's Non-ferrous Metals Industry

1.3.1 Introduction

Mineral resources are the bottleneck of sustainable development of China's mining industry.

On Aug. 30, 2005, the Third China (Western) International Forum for Non-Ferrous Metal Mining Development discussed how to increase the exploration and evaluation work of mineral resources to maintain the dynamic equilibrium of reserves, how to accelerate the establishment of a resource-saving society, how to vigorously develop the circular economy and how to increase the efficiency in the use of mineral resources. Table 1-5 shows the monthly survey of non-ferrous metal ores mining and dressing in China. Table 1-6 gives the investment in urban area by sector, source of funds, jurisdiction of management and registration status (2004).

Table 1-5 Non-ferrous metal ores mining and dressing-monthly survey

Time (year, month)	Number of enterprises, unit	Number of loss-making enterprises, unit	Total assets, 1000 yuan	Ratio of debts to assets, %	Sales revenue of products, 1000 yuan	Total profits, 1000 yuan	Number of employees, person
2004, 02	1, 207	177	56, 289, 182	60	7, 404, 922	696, 958	381, 115
2004, 03	1, 224	179	61, 867, 862	62	13, 201, 440	1, 394, 104	389, 448
2004, 04	1, 224	175	62, 213, 279	61	18, 958, 668	2, 130, 742	391, 858
2004, 05	1, 227	159	63, 622, 341	61	24, 551, 555	2, 848, 278	393, 904
2004, 06	1, 240	162	64, 447, 266	59	31, 299, 134	3, 941, 819	394, 379
2004, 07	1, 240	151	65, 337, 825	58	37, 262, 893	4, 782, 172	395, 878
2004, 08	1, 233	139	63, 525, 823	56	43, 265, 901	5, 858, 648	385, 442
2004, 09	1, 255	143	65, 124, 970	56	50, 758, 686	7, 058, 068	388, 915
2004, 10	1, 262	138	66, 897, 245	55	58, 594, 180	8, 354, 521	391, 081
2004, 11	1, 265	140	69, 716, 277	55	66, 396, 773	9, 572, 826	392, 386
2004, 12	1, 281	129	71, 039, 481	55	74, 061, 546	11, 114, 135	396, 405
2005, 02	1, 239	176	62, 242, 732	53	11, 640, 099	2, 156, 932	357, 414
2005, 03	1, 315	198	64, 201, 126	53	20, 368, 672	3, 729, 422	364, 121
2005, 04	1, 349	213	67, 814, 623	53	28, 513, 755	5, 183, 082	370, 835
2005, 05	1, 367	202	70, 459, 106	52	37, 053, 541	6, 817, 905	378, 702
2005, 06	1, 393	199	72, 740, 141	52	47, 458, 701	8, 913, 423	387, 680
2005, 07	1, 384	194	74, 658, 348	52	56, 340, 516	10, 580, 116	387, 525
2005, 08	1, 386	190	76, 959, 645	51	65, 329, 964	12, 467, 575	389, 637
2005, 09	1, 416	182	79, 390, 057	51	75, 926, 268	14, 591, 932	400, 357
2005, 10	1, 406	180	82, 377, 598	50	85, 171, 407	16, 234, 553	399, 121
2005, 11	1, 421	164	84, 633, 111	50	95, 105, 241	18, 121, 630	399, 408
2005, 12	1, 445	164	86, 310, 543	50	107, 065, 296	19, 912, 577	398, 445
2006, 02	1, 517	247	95, 792, 037	51	17, 236, 170	2, 716, 159	388, 049
2006, 03	1, 553	233	99, 163, 267	50	29, 105, 920	5, 042, 756	395, 153
2006, 04	1, 548	224	104, 084, 906	50	41, 223, 709	7, 512, 121	404, 253
2006, 05	1, 568	220	107, 437, 787	50	54, 663, 065	10, 613, 793	410, 984
2006, 06	1, 574	198	113, 028, 719	50	72, 276, 659	14, 363, 701	414, 142

Continued 1-5

Time (year, month)	Number of enterprises, unit	Number of loss-making enterprises, unit	Total assets, 1000 yuan	Ratio of debts to assets, %	Sales revenue of products, 1000 yuan	Total profits, 1000 yuan	Number of employees, person
2006,07	1,588	195	115,945,699	49	87,638,601	17,628,132	417,965
2006,08	1,610	199	121,659,420	50	102,145,137	21,007,731	422,388
2006,09	1,634	207	126,139,794	49	119,032,688	24,862,195	425,146
2006,10	1,655	204	130,886,543	49	135,191,775	28,252,232	429,818
2006,11	1,677	194	136,151,615	48	151,440,572	31,717,992	432,987
2006,12	1,688	174	143,599,903	48	168,675,454	35,841,338	440,146
2007,02	1,793	221	130,089,363	49	25,753,090	5,048,581	407,297
2007,05	1,905	269	155,675,402	48	84,753,861	17,578,277	477,103
2007,08	1,981	248	176,406,895	46	148,840,361	30,446,381	477,595
2007,11	2,076	242	195,948,784	46	216,030,146	43,318,538	492,197

Table 1-6 Investment in urban area by sector, source of funds, jurisdiction of management and registration status (2004) (Unit: 10^8 yuan)

Item	Investment	Central investment	Local investment	Domestic fund	Fund from Hong Kong, Macao and Taiwan	Foreign fund	State-owned and state-controlled	Collective-owned	Private and individuals
National total	59,028.2	7,339.7	51,688.5	52,060.7	3,113.5	3,854.0	34,091.8	1,879.1	6,517.9
Agriculture, forestry, animal husbandry and fishing	645.1	43.0	602.1	619.5	8.5	17.1	439.5	20.5	70.1
Agriculture	170.6	17.5	153.0	160.0	3.7	6.8	112.1	6.8	18.2
Forestry	145.5	5.4	140.1	143.0	0.6	2.0	123.4	3.8	5.5
Animal husbandry	135.3	4.1	131.2	129.9	1.9	3.4	61.9	5.0	29.4
Fishing	30.9	0.5	30.3	26.0	1.3	3.6	8.0	1.6	9.4
Service activities for agriculture, forestry, animal husbandry and fishing	163.0	15.5	147.5	160.6	1.1	1.2	134.0	3.4	7.6
Mining	2,126.3	1,061.5	1,064.8	2,100.6	15.6	10.1	1,781.2	46.0	121.2
Mining and washing of coal	690.4	67.0	623.4	688.6	1.4	0.4	550.8	25.7	43.2
Extraction of petroleum and natural gas	1,112.3	980.1	132.1	1,100.3	8.8	3.2	1,082.7	1.3	3.1
Mining and processing of ferrous metal ores	131.4	12.0	119.4	129.6	1.4	0.4	49.0	11.2	37.8

Continued 1-6

Item	Investment	Central investment	Local investment	Domestic fund	Fund from Hong Kong, Macao and Taiwan	Foreign fund	State-owned and state-controlled	Collective-owned	Private and individuals
Mining and processing of non-ferrous metal ores	116.7	1.6	115.2	111.0	2.3	3.5	76.2	4.1	12.5
Mining and processing of nonmetal ores	72.8	0.8	72.0	68.4	1.7	2.7	21.0	3.6	23.8
Mining of other ores	2.6		2.6	2.6			1.5	0.2	0.7
Manufacturing	14,657.2	1,432.0	13,225.3	10,796.6	1,238.2	2,622.3	5,867.4	482.1	1,689.5
Processing of food from agricultural products	543.3	5.6	537.7	450.7	34.2	58.4	131.7	21.5	101.0
Manufacture of foods	367.6	4.4	363.2	308.8	24.3	34.5	97.2	14.1	58.3
Manufacture of beverages	254.0	1.4	252.5	211.2	18.4	24.4	95.1	6.9	30.2
Manufacture of tobacco	86.9	64.4	22.5	86.2	0.4	0.3	84.6	0.2	0.0
Manufacture of textile	763.0	1.9	761.1	614.5	79.3	69.2	171.7	29.5	129.5
Manufacture of textile wearing apparel, footwear, and caps	259.0	0.1	259.0	162.4	49.7	47.0	30.0	17.2	54.5
Manufacture of leather, fur, feather and related products	116.4	0.4	116.0	74.5	19.0	22.9	19.7	2.4	20.2
Processing of timber, manufacture of wood, bamboo, rattan, palm, and straw products	178.2	0.1	178.1	147.0	20.5	10.8	40.4	4.8	41.7
Manufacture of furniture	105.3		105.3	65.5	20.5	19.3	15.4	2.8	23.5
Manufacture of paper and paper products	422.7	1.6	421.2	252.1	35.5	135.1	148.5	17.1	42.1
Printing, reproduction of recording media	166.5	10.6	155.9	136.2	22.6	7.6	42.9	10.6	31.8
Manufacture of articles for culture, education and sport activity	54.4		54.4	30.1	10.6	13.6	10.4	1.2	9.1
Processing of petroleum, coking, processing of nuclear fuel	637.9	182.7	455.2	593.2	28.0	16.7	351.0	13.1	65.8
Manufacture of raw chemical materials and chemical products	1,581.9	461.0	1,120.9	990.4	75.7	515.8	949.0	29.3	113.7

Continued 1-6

Item	Investment	Central investment	Local investment	Domestic fund	Fund from Hong Kong, Macao and Taiwan	Foreign fund	State-owned and state-controlled	Collective-owned	Private and individuals
Manufacture of medicines	594.5	3.0	591.5	512.5	42.4	39.5	197.2	21.1	64.0
Manufacture of chemical fibers	171.4	24.9	146.5	105.4	18.3	47.7	59.8	2.9	8.1
Manufacture of rubber	178.6	0.4	178.2	111.4	8.7	58.4	56.3	3.8	20.6
Manufacture of plastics	330.9	3.0	327.9	210.1	44.4	76.5	52.6	8.9	60.2
Manufacture of non-metallic mineral products	1,101.8	11.4	1,090.5	936.1	90.2	75.5	342.2	53.4	196.4
Smelting and pressing of ferrous metals	1,789.6	208.1	1,581.5	1,582.6	72.8	134.1	1,176.9	36.4	90.3
Smelting and pressing of non-ferrous metals	572.7	92.2	480.4	493.7	59.8	19.1	334.7	18.7	42.4
Manufacture of metal products	430.6	5.5	425.1	320.3	46.2	64.1	73.6	23.3	99.9
Manufacture of general purpose machinery	562.0	20.4	541.5	462.3	29.6	70.1	190.7	31.9	98.6
Manufacture of special purpose machinery	464.0	47.3	416.8	377.1	28.1	58.8	163.5	20.9	69.4
Manufacture of transport equipment	1,033.2	254.6	778.6	654.3	45.8	333.0	597.9	16.7	60.3
Manufacture of electrical machinery and equipment	525.7	6.4	519.3	372.9	55.7	97.1	118.4	33.1	82.3
Manufacture of communication equipment, computers and other electronic equipment	1,021.2	14.4	1,006.8	278.8	221.9	520.4	195.2	18.9	30.5
Manufacture of measuring instruments and machinery for cultural activity and office work	130.7	5.4	125.3	78.5	21.6	30.6	34.2	3.6	10.7
Manufacture of artwork and other manufacturing	200.5	0.9	199.5	168.1	13.2	19.1	81.5	16.1	33.2
Recycling and disposal of waste	12.9		12.9	9.5	0.6	2.8	4.9	1.6	1.2
Production and distribution of electricity, gas and water	5,525.1	1,733.0	3,792.1	5,079.2	261.8	184.1	4,480.0	75.9	101.4

Continued 1-6

Item	Investment	Central investment	Local investment	Domestic fund	Fund from Hong Kong, Macao and Taiwan	Foreign fund	State-owned and state-controlled	Collective-owned	Private and individuals
Production and distribution of electric power and heat power	4,854.4	1,699.1	3,155.3	4,462.9	237.5	154.0	3,924.5	63.0	87.5
Production and distribution of gas	209.8	32.0	177.8	170.7	14.0	25.1	141.4	3.4	5.9
Production and distribution of water	460.9	1.9	459.0	445.6	10.3	5.0	414.2	9.6	8.0
Construction	526.3	37.6	488.8	516.3	5.9	4.1	366.6	41.6	34.6
Construction of buildings and civil engineering	461.8	34.1	427.6	453.8	5.0	3.0	330.8	36.5	26.7
Building installation	18.6	3.4	15.2	18.0		0.5	7.9	2.2	3.5
Building decoration	8.0		8.0	7.7	0.3	0.1	0.7	0.9	2.2
Other construction	38.0		38.0	36.9	0.6	0.5	27.2	2.0	2.1
Transport, storage and post	7,091.5	1,371.5	5,720.0	6,958.2	68.9	64.4	6,627.9	59.4	50.7
Railway transport	846.3	752.7	93.6	846.3			837.8	4.2	0.4
Road transport	4,665.5	200.7	4,464.8	4,626.5	26.9	12.1	4,442.1	33.3	19.8
Urban public transport	391.3	0.2	391.1	387.0	2.2	2.1	331.4	3.9	2.0
Water transport	534.6	106.6	428.1	492.3	27.2	15.1	479.0	1.6	9.8
Air transport	272.4	166.6	105.8	270.7	1.7		267.5	0.4	
Transport via pipelines	107.0	88.2	18.9	101.7	3.3	2.0	99.9	4.4	0.1
Loading, unloading and other transport services	58.6	0.1	58.5	57.9	0.6	0.1	54.3	0.4	1.9
Storage	186.9	37.3	149.6	147.0	6.9	33.0	89.0	11.2	14.9
Post	28.8	19.2	9.6	28.8			27.0		1.7
Information transmission, computer services and software	1,638.0	776.2	861.8	1,221.7	247.8	168.4	1,297.4	8.4	4.2
Telecommunications and other information transmission services	1,589.9	770.4	819.5	1,177.2	246.0	166.7	1,273.3	4.9	0.8
Computer services	11.7	1.7	10.1	10.8	0.4	0.5	6.9		0.6
Software	36.4	4.1	32.3	33.8	1.3	1.3	17.1	3.5	2.8

Continued 1-6

Item	Investment	Central investment	Local investment	Domestic fund	Fund from Hong Kong, Macao and Taiwan	Foreign fund	State-owned and state-controlled	Collective-owned	Private and individuals
Wholesale and retail trade	1,117.2	30.6	1,086.6	1,048.1	35.3	33.8	368.5	99.4	176.2
Wholesale trade	447.4	23.5	423.8	420.8	18.3	8.3	162.2	34.8	63.3
Retail trade	669.8	7.0	662.8	627.3	17.0	25.5	206.3	64.6	112.9
Hotel and restaurants	438.1	5.7	432.4	367.0	42.6	28.5	124.2	18.1	95.9
Hotels	295.0	5.5	289.5	237.5	34.3	23.2	100.2	10.1	45.1
Restaurants	143.1	0.2	142.9	129.5	8.3	5.3	24.0	8.0	50.8
Financial intermediation	97.6	54.0	43.6	94.9	2.6	0.1	79.5	9.5	0.5
Banks	83.4	50.5	32.9	80.8	2.5		68.7	8.8	0.1
Security activities	1.3	0.1	1.2	1.2		0.1	0.4		0.1
Insurance	5.0	3.4	1.6	4.9			4.2		
Other financial activities	8.0	0.1	7.9	7.9			6.3	0.7	0.4
Real estate	14,547.0	9.1	14,537.9	12,784.6	1,110.6	651.7	3,265.3	696.8	3,972.8
Leasing and business services	361.7	3.5	358.2	348.5	7.3	5.9	193.0	31.7	26.3
Leasing	6.2		6.2	6.1			0.3	0.8	1.1
Business services	355.5	3.5	352.0	342.3	7.3	5.9	192.7	30.9	25.2
Scientific research, technical service and geologic prospecting	311.8	102.6	209.1	301.3	5.2	5.3	268.5	2.8	9.9
Research and experimental development	169.2	73.3	95.8	163.5	1.2	4.5	153.6	0.5	3.3
Professional technical services	71.8	13.4	58.4	68.9	2.6	0.2	59.3	1.2	1.3
Services of science and technology exchanges and promotion	45.5	1.6	43.9	43.6	1.3	0.5	31.4	0.9	4.8
Geologic prospecting	25.3	14.3	11.0	25.3		0.1	24.2	0.2	0.5
Management of water conservancy, environment and public facilities	4,890.8	110.0	4,780.9	4,843.7	15.5	31.6	4,546.1	65.8	31.5
Management of water conservancy	749.6	92.4	657.2	745.5	1.0	3.1	732.7	2.9	1.1

Continued 1-6

Item	Investment	Central investment	Local investment	Domestic fund	Fund from Hong Kong, Macao and Taiwan	Foreign fund	State-owned and state-controlled	Collective-owned	Private and individuals
Environmental management	258.9	3.6	255.3	253.9	0.7	4.3	232.5	2.3	2.1
Management of public facilities	3,882.3	13.9	3,868.4	3,844.3	13.8	24.2	3,580.9	60.6	28.3
Services to households and other services	107.6	1.6	106.0	104.5	0.8	2.2	49.0	6.4	17.0
Services to households	67.3	0.4	66.9	65.7	0.8	0.8	38.2	4.5	10.7
Other services	40.3	1.2	39.1	38.8	0.1	1.5	10.8	1.9	6.3
Education	1,803.0	213.3	1,589.7	1,792.4	5.5	5.0	1,619.1	30.8	64.0
Health, social security and social welfare	446.9	25.2	421.7	440.4	5.0	1.5	400.1	12.0	15.9
Health	419.3	24.9	394.5	412.9	4.9	1.5	375.6	11.2	15.3
Social security	4.5	0.1	4.4	4.5			4.2		
Social welfare	23.1	0.3	22.8	23.1			20.4	0.8	0.6
Culture, sports and entertainment	531.2	25.9	505.3	485.1	30.0	16.0	410.0	14.2	20.3
Journalism and publishing activities	18.7	1.7	16.9	18.7			17.7	0.2	
Broadcasting, movies, televisions and audiovisual activities	74.2	3.4	70.8	74.2	0.1		68.8	0.2	0.7
Cultural and art activities	154.8	9.8	145.0	153.3	0.9	0.6	143.0	1.6	2.9
Sports activities	125.5	8.3	117.1	118.8	1.0	5.7	113.4	1.1	0.4
Entertainment	158.0	2.6	155.4	120.3	28.0	9.8	67.1	11.1	16.3
Public management and social organization	2,165.6	303.4	1,862.2	2,157.7	6.4	1.5	1,908.4	157.7	16.0
Organs of Communist Party of China	30.0	7.2	22.8	30.0			29.3	0.7	
Government agencies	1,907.4	295.2	1,612.3	1,900.1	6.0	1.3	1,818.3	43.6	6.0
People's political consultative conference and democratic parties	2.7		2.7	2.7			2.7		

Continued 1-6

Item	Investment	Central investment	Local investment	Domestic fund	Fund from Hong Kong, Macao and Taiwan	Foreign fund	State-owned and state-controlled	Collective-owned	Private and individuals
Non-governmental institutions, social organizations and religion organizations	29.8	0.9	28.9	29.6	0.1	0.1	16.4	1.4	0.6
Grass roots self-governing organizations	195.7	0.2	195.5	195.3	0.3	0.1	41.7	112.0	9.4
International organizations	0.3		0.3	0.3			0.1		

It is released by China National Development and Reform Commission that China's industrial added value of enterprises above designated size in 2009 increased by 11% year on year (YOY). The increases for the quarters were 5.1%, 9.1%, 12.4%, and 18%, respectively, up quarter by quarter. The production of high energy consumption products such as iron and steel, nonferrous metals, cement, etc. recovered quickly. The prices of steel, cement and most of the chemical products fluctuated slightly, while prices of most nonferrous metals and flat glass increased.

Table 1-7 shows output of key nonferrous metals in 2009. Prices of the main nonferrous metals increased gradually. In December 2009, the spot prices of copper, electrolytic aluminum, lead, and zinc in China were CNY 56,054, CNY 15,992, CNY 15,986, and CNY 18,624 per metric ton, increasing by 7.9%, 5.8%, 0.8%, and 11.2% month on month, and up 110%, 43.4%, 52.9%, and 93.2% year on year(YOY).

Table 1-7 Output of key nonferrous metals

Item	Output, metric tons	YOY, %
Ten non-ferrous metals	27.24×10^6	5.8
Electrolytic aluminum	13.20×10^6	1
Copper	—	9.6
Lead	—	16.4
Zinc	—	11.6
Alumina	—	4.4

1.3.2 Current Situation and Future Outlook of China's Non-ferrous Metals Industry

1.3.2.1 Current Situation

The global production of ferrous and non-ferrous castings is estimated at 81.28 million metric tons and its value exceeds USD 100 billion. The global trade on metal castings is estimated at USD

10 billion and is expected to multiply in the coming years. Non-ferrous metals regularly occur in smelter by-products, in the residues of industrial treatment processes, and also in a variety of complex metal combinations. The processor must treat each of these as quite separate materials, employing skill and technical resources to identify the individual metals and to convert them into a usable form for consuming industries. With the rapid development of the production base in China, India and the Far East, the flow of non-ferrous scrap metals has switched increasingly from West to East.

Recently the attention on the rising price of steel has masked the relevance of the non-ferrous sector aluminium, copper, zinc and lead which plays an equally important role in user industries such as consumer goods, utility and the transport sectors. Non-ferrous metals account for about 2 percent to 3 percent of the total inflation basket. However, this is expected to rise, say industry experts, mainly due to the growing applications of aluminium and copper and also due to an expected rise in prices of these commodities. Globally, non-ferrous metal prices have risen by more than 300 percent over the past four years, mainly due to falling supplies from mines as companies haven' t increased investment in new mining, smelting and processing capacities. Inadequate supplies from mines are not able to keep up demand from emerging markets such as India and other BRIC countries.

In 2006, China made certain progress in upgrading non-ferrous metals industry, and the growth rate of fixed assets investment in non-ferrous metals declined somewhat. The investment in non-ferrous metals industry added up to RMB 1180. 8 billion, up 24. 1% and the growth rate decreased by 11% over 2005-06; besides, the investment in non-ferrous metals industry accounted for 12. 7% of China' s total investment in fixed assets in urban and rural areas. Smelting industry still attracted the largest proportion, yet the newly-started projects in smelting industry was restrained to some extent, and the investment began to transfer towards mines with unsubstantial capital. The investment structure was further improved.

China' s non-ferrous metal price in 2006 continued a large rise after hiking for three consecutive years. Main non-ferrous metals experienced a considerable price hike, especially zinc, copper and nickel. In international spot market, the average prices of zinc, copper and nickel are respectively USD 3, 223/t, USD 6, 606/t and USD 23, 862/t, up by 13. 71%, 8. 25% and 6. 46% YOY. The year 2006 also witnessed the historical highest profit of China' s non-ferrous metals industry. In the whole year, non-ferrous metals enterprises above designated size generated the industrial added value of RMB 358 billion, up 23% over 2005 calculated by comparable prices.

In 2007, China' s nonferrous metals industry was in a period of high prosperity. Consumption grew enormously. This benefited from the support of thriving consumption under a high economic growth rate and also from the industry integration promoted by relevant industry policies. However, weak awareness of brand and insufficient supply has not been improved. The non-renewal ability and scarcity of nonferrous metal resources pushed China' s nonferrous metals market towards concentration and integration. In terms of market structure, copper and aluminum are still the biggest part of China' s nonferrous metals consumption.

At present, China' s non-ferrous metals industry has a momentum of a slowdown growth after years of rapid growth. The reasons for the slower growth in the industry are as follows. Firstly, China' s non-ferrous metal production and consumption accounts for about 30% of the global total, meaning that per capita consumption in China has surpassed the global average level. Secondly, cost continues to rise, and corporate profitability is declining after earlier rises, which gradually dampens the corporate enthusiasm for sustained fast growth in capacity expansion. Thirdly, extensive economic growth has led to a tight supply of resources and energy in China and China has continuously rolled out regulatory and control policies, which will help the industry change the growth mode with focus on capacity expansion, and step onto the path of sustained development.

Fig. 1-2 shows China' s non-ferrous metal output, 1H2007 and 1H2008 (Source: Research in China). It is forecast that China' s output of non-ferrous metals will grow at a rate of around 10% in the period 2008—2010, even lower than 10%, and meanwhile growth rate of consumption will slow down accordingly. After years of rapid rises, non-ferrous metal prices have begun dividing. Some of non-ferrous metal prices have dropped due to the impact of supply growth, and some have still remained at high level because of tight supply. With the rapid development of non-ferrous metal industry in recent years, large-sized corporations have taken the initiative in the rapid growth. Aluminum Corporation of China, for instance, will dominate the future development of China electrolyte aluminum industry. The four gold producers, Zhongjin Gold Co, Shandong Gold Co, Zhaojin Industry and Zijin Mining Group, will share the gold market, while the copper market will be controlled by Tongling Nonferrous Metals Group Holdings Co., Jiangxi Copper Industry Co. and Yunnan Copper Industry Co. China' s investment in overseas market by its non-ferrous metal producers is increasingly active, but they are still at initial stage and still face great risk. In a long term, it is an inevitable trend for China' s non-ferrous metal companies to tap overseas market and participant in global competition in the future.

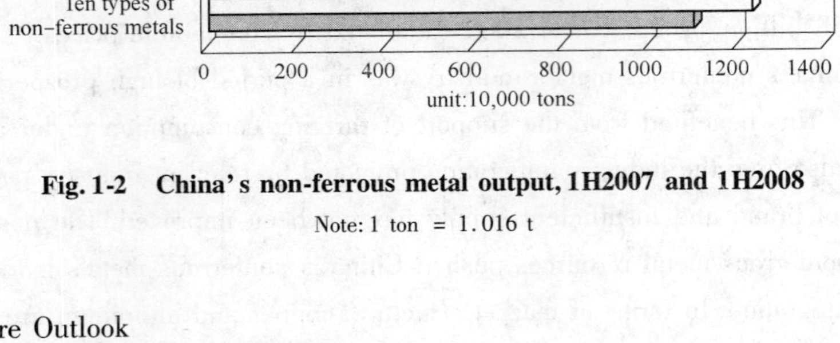

Fig. 1-2 China' s non-ferrous metal output, 1H2007 and 1H2008

Note: 1 ton = 1.016 t

1.3.2.2 Future Outlook

(1) According to recent media reports, global aluminium prices are expected to rise by addi-

tional 50 percent through 2009 due to reduction of output by China' s top aluminium producers. This is part of a Chinese government-controlled measure to reduce consumption of electricity as the country' s power supplies have fallen short.

(2) Electricity accounts for three-fourth of the total cost of aluminium. According to the China Non-Ferrous Metals Industry Association, the Aluminum Corp of China, the largest producer, and a host of other aluminium producers which account for about 70 percent of the total aluminium production of China, have agreed to reduce production by 406, 400 metric tons. This would save 5. 8 billion kilowatt-hours of electricity.

(3) The impact from China' s measures have already started with the monthly sales of aluminium on the London Metal Exchange rising by about 40 percent, with analysts predicting a further rise once China becomes a net importer of aluminium due to the reduction in output.

(4) Although domestic availability of aluminium is not a problem——about 1 million tonnes is produced in India——the high prices and slow progress in the expansion projects of Sterlite Industries and Hindalco Industries, could bring little comfort to prices.

(5) In copper, the local situation is no different either. Hindalco Industries recently said that a scoping study at one of its mines in Australia has concluded that the mine is commercially unviable. Although the Aditya Birla group has another large copper mine at Nifty in Australia, the scoping study can dampen sentiment.

(6) Global copper cathode market would remain tight because of a severe shortage in concentrates the main raw material for making copper. Current global inventories account for less than three weeks worth of world consumption.

(7) The bleak outlook on the concentrate front has also prompted the government to mull revival of a closed mine of the state-run Hindustan Copper, in Rajasthan. Although the issue is still being discussed, senior company officials have said that the government has been seriously considering the mine revival as it feels that metal prices are fast becoming a vital part of inflation.

(8) Global copper mine production will grow by only 355, 600 metric tons this year, while consumption will grow by nearly 750, 000 metric tons due to demand from Latin America, the Middle East, China and other parts of Asia.

Professional Terms

geology 地质(学)
mining 采矿,矿业
dressing(mineral processing) 选矿
metallurgy 冶金
environment 环境
exploration 勘探
mineral 矿物
rock 岩石
ore 矿石
RE (rare earths) 稀土
mineral resource 矿物资源
mineral conservation 矿物资源保护
non-ferrous metal deposit 有色金属矿床
non-ferrous metals industry 有色金属工业

Exercises

1-1 简述采矿的地位与作用。

1-2 简述采矿对环境的影响。

1-3 常用选矿方法有哪些？请举几个联合选矿方法的例子。

1-4 简述从矿床勘探、采矿、选矿、冶炼、材料加工到产品销售的整个活动过程。

1-5 简述中国有色金属工业的现状与未来。

1-6 “BRIC countries”是指哪些国家？

1-7 “1H2008”是什么意思？

1-8 分析下列句子的语法结构：

(1) Since most of the useful elements compose such a small percentage of the earth' s crust, the occurrence of ore deposits as we know them would not have transpired had not geologic processes concentrated the elements. (See Section 1. 1. 1 paragraph 4)

(2) The earliest metal implement found that has been dated was one in Egypt from 3700 B. C. Bronze was used in Crete from 3400 B. C. (See Section 1. 1. 3 paragraph 6)

2

Explosives and Blasting

本章介绍炸药与爆破的有关知识，包括炸药的性质（如猛度、爆速、密度、爆炸压力、炮孔压力、抗水性和感度）、种类（如硝化甘油炸药、硝铵炸药、胶质炸药和浆状炸药）、价格和选用，爆破器材（如雷管、导爆索、起爆药包和传爆药包），炮孔装药方法及控制爆破技术（如预裂爆破、缓冲爆破和光面爆破）。

Blasting technology has become very dynamic. The development of low-cost blasting agents and more efficient loading methods have continually reduced blasting costs while other excavation costs steadily rise. The most significant development in blasting products was ammonium nitrate-fuel oil (ANFO) blasting agents which, introduced in the mid-1950s, accounted for over 75%, by weight, of the 1970 market. Slurries (water gels) first appeared commercially in 1959 and account for about 10% of the consumption. Nitroglycerin (NG) based explosives, once the standard of the industry, represent only about 15% of the market. Consumption trends are fairly stable. Slurry use is gradually growing and NG explosive use is gradually diminishing. Loading techniques have evolved concurrently with blasting products. Individual loading and tamping of cartridged high explosives is being largely replaced by pneumatic and auger loading of dry blasting agents, and pumping and extruding of slurry products.

This chapter deals primarily with explosive types, their properties and methods of blasthole loading. At times, the term "explosives" will be used collectively to describe all products that will detonate. The differences between high explosives and blasting agents will be described later.

2.1 Characteristics of an Explosive Reaction

An explosive or blasting agent is a compound, or a mixture of compounds, which, when initiated by heat, impact, friction or shock, undergoes a very rapid self-propagating exothermic reaction.

This reaction produces more stable products, usually gases, which exert tremendous pressure as they expand at high temperature. The term "detonation" indicates that the reaction is moving through the explosive faster than the local speed of sound in the unreacted explosive, creating a shock wave, whereas "deflagration" indicates a reaction slower than the speed of sound. A high explosive will detonate. A low explosive, such as black powder, normally will deflagrate. The work done by an explosive depends on both the shock energy and the energy of the expanding gas produced by the explosion.

For mixtures of carbon, hydrogen, oxygen and nitrogen compounds, energy release is maximized at zero oxygen balance, the point at which there is sufficient oxygen to completely oxidize all the fuels in the mixture, but no excess to react with the nitrogen in the mixture to form nitrogen oxides. Theoretically, at zero oxygen balance, the gases produced upon detonation are H_2O, CO_2 and N_2, although in reality small amounts of NO, CO, NH_3, CH_4 and other gases are generated. Solid detonation products, usually metallic oxides, are produced when the explosive contains such ingredients as sodium nitrate or aluminum.

The reaction equations for ANFO, the simplest blasting-agent mixture, describe the relationship of oxygen balance, detonation products and heat release. The equations assume an ideal detonation reaction, which in turn assumes thorough mixing of ingredients, proper particle sizing, adequate confinement, charge diameter and priming, and protection from water. Fuel oil is not precisely CH_2, but this identification simplifies the equations and is accurate enough for the purposes here.

These equations illustrate simplified reactions between oxidizers (ammonium nitrate) and fuel-sensitizers (fuel oil and aluminum). Explosives often contain such additional ingredients as antacids, absorbents or stabilizers, which enter into the explosive reaction. Paper cartridges, paraffin coatings and plastic borehole liners are at least partially consumed by the detonation. The resulting explosive reactions are much more complex than those described here.

$$3NH_4NO_3(94.5\%) + CH_2(5.5\%) \longrightarrow 7H_2O + CO_2 + 3N_2 + \approx 0.94\text{kcal per g} \quad (2\text{-}1)$$

$$2NH_4NO_3(92.0\%) + CH_2(8.0\%) \longrightarrow 5H_2O + CO + 2N_2 + \approx 0.82\text{kcal per g} \quad (2\text{-}2)$$

$$5NH_4NO_3(96.6\%) + CH_2(3.4\%) \longrightarrow 11H_2O + CO_2 + 4N_2 + 2NO + \approx 0.61\text{kcal per g} \quad (2\text{-}3)$$

Eq. 2-1 represents the ideal reaction of an oxygen-balanced mixture containing 94.5% AN and 5.5% FO. None of the detonation products are poisonous, and about 0.94 kcal of heat are released for each gram of ANFO detonated. In Eq. 2-2, a mixture of 92.0% AN and 8.0% FO creates an oxygen deficiency. As a result, the carbon in the fuel oil is oxidized only to CO, a poisonous gas, rather than relatively harmless CO_2. Because of the lower heat of formation of CO, only about 0.82 kcal of heat is released for each gram of ANFO detonated. In Eq. 2-3 the mixture of 96.6% AN and 3.4% FO has a fuel shortage, which creates an excess oxygen condition. Some of the nitrogen from the ammonium nitrate combines with this excess oxygen to form nitrous oxide (NO) which, upon exposure to a normal atmosphere, forms NO_2 an extremely toxic gas. An oxygen-balanced mixture, then, maximizes energy release while minimizing formation of toxic gases.

Although the simple ANFO mixture is optimum for highest energy release per unit cost of ingredients, products with higher energies and densities often are desired. The effect of adding aluminum to an ANFO mixture is shown in Eq. 2-4:

$$6NH_4NO_3(87.6\%) + CH_2(2.5\%) + 2Al(9.9\%) \longrightarrow$$
$$Al_2O_3 + 13H_2O + CO_2 + 6N_3 + \approx 1.32\text{kcal per g} \quad (2\text{-}4)$$

The higher heat of explosion is due to the high heat of formation of Al_2O_3. The increased density and increased energy per unit weight of an aluminized blasting agent must be weighed against its higher cost.

For further illustration it will be helpful to take a brief look at the detonation process, as pictured by R. L. Ash, of the University of Missouri-Rolla. Fig. 2-1, adapted from Ash' s work, depicts a column of slurry explosive or slurry blasting agent which has been initiated at the left. Detonation has proceeded to the center of the column. The primary reaction occurs between the shock front and a rear boundary known as the Chapman-Jouget (C-J) plane. The length of this primary reaction zone is an inherent characteristic of each explosive substance or mixture. The critical diameter, or minimum diameter at which an explosive reaction will propagate dependably, is a function of the length of the reaction zone, which depends on the type of ingredients, particle size, density, and confinement. Because military explosives, such as PETN or TNT, have very short reaction zones, their critical diameters are quite small. Cartridged explosives have somewhat longer reaction zones, and blasting agents and slurries have still longer. Consequently, their critical diameters are larger than those of dynamites and gelatins. As these critical diameters are approached, the detonation velocity and blasting performance of blasting agents and slurries decrease markedly.

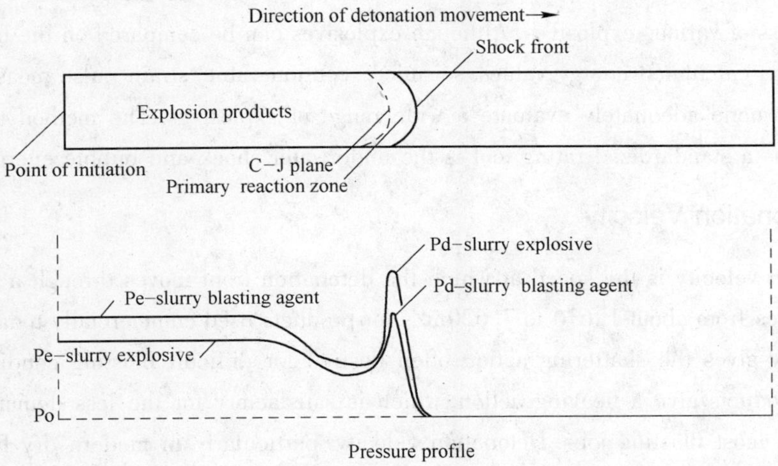

Fig. 2-1 Simplified illustration of detonation reaction

The pressure profiles in Fig. 2-1 show the explosive forces applied to the rock being blasted. The figure gives a general comparison between an explosive slurry and a blasting agent. The rock is first hit by the supersonic blow from the shock front, characterized by the detonation pressure, Pd.

It is this detonation pressure that gives an explosive its shattering action. This is followed by a brief release of pressure which in turn is followed by a prompt buildup of pressure caused by the explosion products. The latter effect is called borehole or explosion pressure, Pe, which gives the explosive its heaving action. In high explosives, since most of the reaction of the ingredients is completed in the primary reaction zone, detonation pressures are high. The ensuing borehole pressure depends on the amount of gas generated and the heat of explosion. In blasting agents and slurries, significant reactions occur behind the C-J plane. These include the reaction of larger ingredient particles and the oxidization of slowly reacting metallic fuels, such as aluminum. As a result of these delayed reactions, detonation pressures of blasting agents generally are lower than those for high explosives because more of the energy is released in the later explosion phase. The resulting explosion pressures are, however, similar to those of some high explosives. It is important to keep these points in mind when comparing explosives with blasting agents. Based on detonation pressure alone, high explosives will attain higher ratings, but in overall performance blasting agents are quite competitive.

2.2 Comparative Properties of Explosives

Explosives can be classified according to a number of properties. Depending on specific applications, a given property may be of critical or very little importance.

2.2.1 Strength

The terms "weight strength" and "cartridge strength", which originally indicated the percentage of nitroglycerin in an explosive, were useful when it was the principal energy-producing ingredient. However, with development of products with decreasing proportions of nitroglycerin, these strength ratings have become misleading and inaccurate and should be disregarded when comparing the effectiveness of various explosives. Although explosives can be compared on the basis of several systems, such as calculated energy values, seismic execution value, strain pulse measurements, and crater methods, none adequately evaluate a wide range of explosives. The method that shows the most promise as a standardized rating tool is the underwater shock and bubble energy test.

2.2.2 Detonation Velocity

Detonation velocity is the speed at which the detonation front moves through a column of explosive. It ranges from about 1,676 to 7,620m/s for products used commercially today. A high detonation velocity gives the shattering action often desired for difficult blasting conditions, whereas low-velocity products give a heaving action, which is satisfactory for the less demanding requirement typical of most blasting jobs. Detonation velocity, particularly in modern dry blasting agents and slurries, may vary widely depending on field conditions. Detonation velocity can often be increased by the following:

(1) Using a larger charge diameter.

(2) Increasing density (although excessively high densities in blasting agents may seriously reduce sensitivity).

(3) Decreasing particle size (pneumatic injection of ANFO in small-diameter boreholes illustrates this).

(4) Providing good confinement in the borehole.

(5) Providing a high coupling ratio (coupling ratio is the percentage of the borehole diameter filled with explosive).

(6) Using a larger initiator or primer.

2.2.3 Density

Density normally is expressed in terms of specific gravity, the ratio of the density of the explosive to that of water. A useful form of density is loading density, the weight of explosive per unit length of charge, commonly expressed in kilograms per meter. Fig. 2-2 is a nomograph for determining loading density. Cartridge count (number of 32 ×203mm cartridges per 22.65kg box) is useful when dealing with cartridged high explosives and is approximately equal to 141 divided by the specific gravity. The specific gravity of commercial products ranges from 0.5 to 1.7. The density of an explosive determines the weight that can be loaded into a given volume of borehole, and is particularly important where drilling is expensive. The energy per unit volume of explosive, however, actually is a more important consideration, although not a commonly reported explosive property.

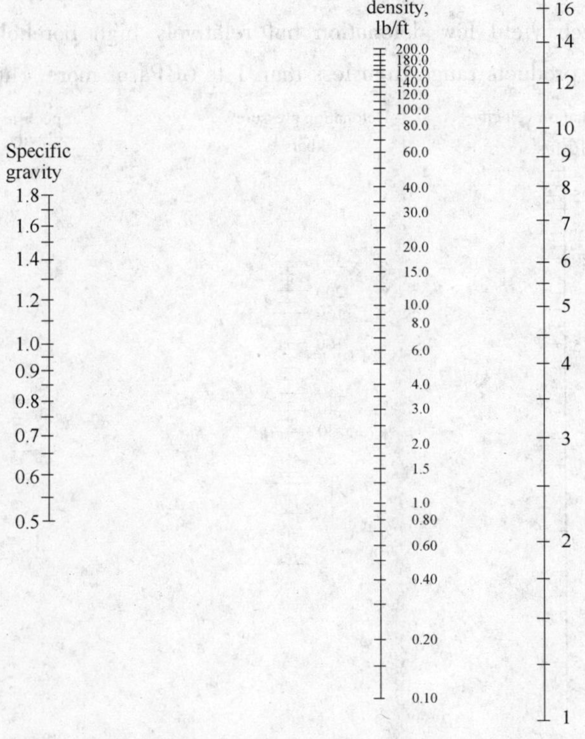

Fig. 2-2 Nomograph for determining loading density

Note: 1in = 25.4mm; 1lb = 0.4536kg; 1ft = 0.3048m

2. 2. 4 Detonation Pressure and Borehole Pressure

These are two key properties of an explosive, although authorities differ on the relative importance of the two in breaking rock. Detonation pressure is the pressure of the detonation wave propagating through the explosive column, measured at the C-J plane (Fig. 2-1). Although the relationship of detonation velocity and density to detonation pressure is somewhat complex, and depends on the ingredients of an explosive, the following approximation is one of several that can be made:

$$P = 3.88 \times 10^{-16} DC^2/(1 + 0.80D) \tag{2-5}$$

where P is detonation pressure, Pa ($1\text{Pa} = 1\text{N/m}^2$), D is specific gravity and C is detonation velocity, m/s.

The nomograph in Fig. 2-3, based on this formula, can be used to approximate the detonation pressure of an explosive when the detonation velocity and specific gravity are known. It generally is agreed that a high detonation pressure resulting in a strong shock wave is of major importance in breaking very dense competent rock. Detonation pressures for commercial products range from about 0.5 to over 15GPa. Borehole pressure, also called explosion pressure, is the pressure exerted on the borehole walls by the expanding gases of detonation after the chemical reaction has been completed. It is a function of confinement and the quantity and temperature of the gases of detonation. Borehole pressure generally is considered to be dominant in breaking softer or weaker rocks and in displacing all types of rock encountered in blasting. This characteristic accounts for the success of ANFO and aluminized products, which yield low detonation but relatively high borehole pressures. Borehole pressures for commercial products range from less than 1 to 6GPa or more. They are calculated from

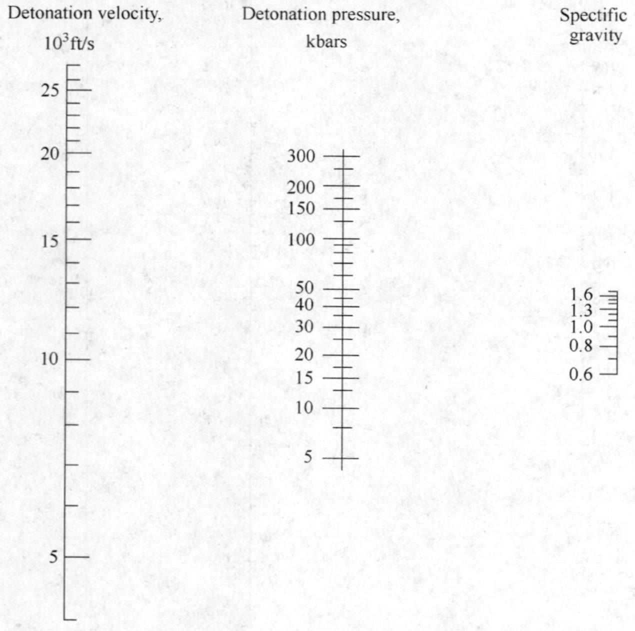

Fig. 2-3 Nomograph for finding detonation pressure

Note: $1\text{ft/s} = 0.3048\text{m/s}$; $1\text{kbar} = 1\text{kb} = 1000\text{b} = 10^8\text{Pa} = 10^8\text{N/m}^2$

hydrodynamic codes or approximated from underwater test results, since borehole pressure cannot, at present, be measured directly. Many ANFO mixtures have borehole pressures larger than their detonation pressures. In most high explosives, the detonation pressure is the greater.

2.2.5 Water Resistance

Water resistance is ability to withstand exposure to water without losing sensitivity or efficiency. Gelled products, such as gelatin dynamites and slurries, have good water resistance. Nongelatinized high explosives have poor to good water resistance. Dry blasting agents, such as ANFO, have practically no water resistance. Ammonium nitrate prills should not be used in the presence of excess water. The emission of brown nitrogen oxide fumes from a blast often indicates inefficient detonation caused by water deterioration, and signifies the need for a more water-resistant explosive or external protection from water in the form of a plastic sleeve. Certain precautions in the use of plastic sleeves will be noted later.

2.2.6 Fume Class

Fume class is a measure of the amount of toxic gases, primarily carbon monoxide and oxides of nitrogen, produced by the detonation of an explosive. Most modern blasting products are oxygen-balanced both to minimize fumes and to optimize energy release per unit cost of ingredients. Fumes are an important consideration in tunnels, shafts and other confined spaces. Certain blasting conditions may produce toxic fumes even with oxygen-balanced explosives. Insufficient charge diameter, inadequate priming or initiation, water deterioration, removal of wrappers or the use of plastic borehole liners all increase the likelihood of generating toxic gases. The following table shows fume classes adopted by the Institute of Makers of Explosives (Bichel-gage method), U. S. Bureau of Mines standards limit the volume of poisonous gases produced by a permissible explosive to about 0.156 cu. m per kg of explosive.

Fume class	Cu. m of poisonous gases per 200 g of explosive
1	<0.00453
2	0.00453—0.00934
3	0.00934—0.20422

2.2.7 Sensitivity

Sensitivity of explosives has been expressed in many ways. Some, such as sensitivity to friction, heat and impact, are important safety considerations when handling the more sensitive dynamites and detonators. Air-gap sensitivity is important when using cartridged products where the possibility of charge separation exists. A card gap test, in which a nonexplosive substance is interposed between the donor and acceptor charges, is a basic sensitivity measurement for an explosive. In this age of insensitive dry blasting agents and slurries, critical diameter and minimum primer requirement have become the useful measures of "performance sensitivity." Critical diameter is the minimum diameter at which a product will propagate dependably. Manufacturers recommend as a minimum diameter the critical diameter plus a safety factor. In all commercial products, particularly dry blasting agents and slurries, the detonation velocity and the ability to break rock are reduced as the critical diameter is approached. Fig. 2-4, taken from R. L. Ash, University of Missouri-Rolla, illustrates this effect for various explosive products. Critical diameters can vary from less than 12. 7mm for a specially sensitized slurry to nearly 25. 4mm for a finely pulverized ANFO product, and can be 50. 8mm or greater for some of the less sensitive blasting agents. Minimum primer requirement measures a product' s susceptibility to initiation and can vary from a No. 6 cap for most high explosives to much stronger sources for some blasting agents. Since minimum primer requirement is not a precise value, some degree of overpriming is considered inexpensive insurance for good blasting-agent performance. Products with low primer requirements usually have small critical diameters.

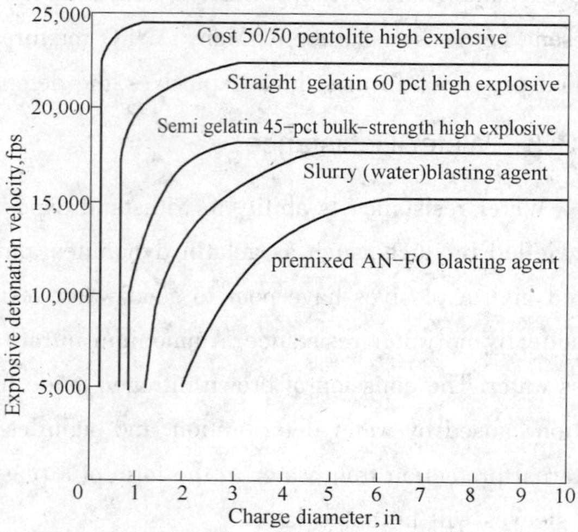

Fig. 2-4 The effects of charge diameter on detonation velocity

Note: 1in = 25. 4mm; 1fps = 0. 3048m/s

2.3 Commercial Explosive Products

Products used for the main charge in a blasthole can be divided into three categories: nitroglycerin-based high explosives, including permissibles; dry blasting agents, including rigidly cartridged blasting agents (RCBA); and slurries or water gels. Fig. 2-5 shows the relative consumption of these products in the United States in 1970. Fig. 2-6 gives the breakdown by industry.

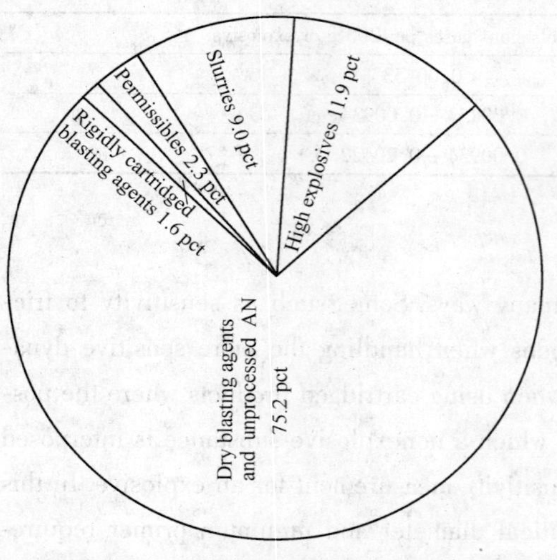

Fig. 2-5 Consumption of blasting agents and explosives in the United States in 1970

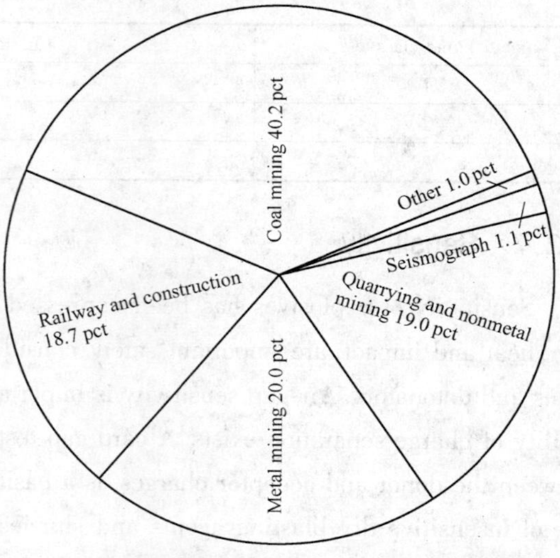

Fig. 2-6 Consumption of blasting agents and explosives in the United States in 1970, by industry

2. 3. 1 Nitroglycerin-Based Explosives

Nitroglycerin-based explosives can be further categorized as to their nitroglycerin content. Fig. 2-7 shows this breakdown along with some relative properties and ingredients of these products. Table 2-1 shows the properties of nitroglycerin-based explosives. Property values are averages of manufacturers' published figures.

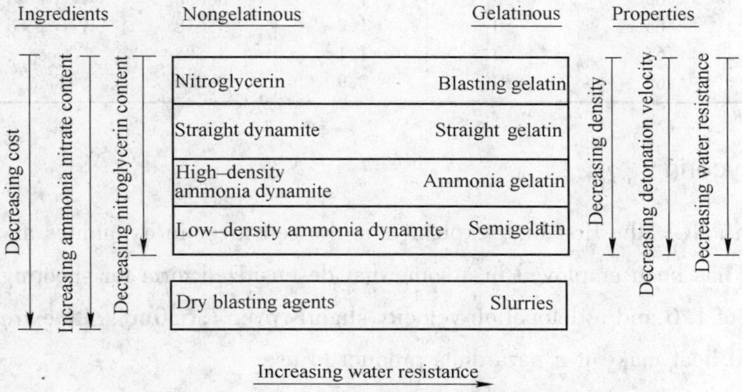

Fig. 2-7 Some relative properties and ingredients of commercial explosives and blasting Agents

Table 2-1 Properties of NG-based explosives

Type	Weight strength, %	Bulk strength, %	Specific gravity	Detonation velocity, m/s	Water resistance	Fume class
Straight dynamite	60	60	1.3	5,791	Good	Poor
	40	40	1.4	4,267	Good	Poor
	20	20	1.4	2,743	Poor	Poor
High-density ammonia dynamite	60	50	1.3	3,810	Fair	Good
	40	35	1.3	3,200	Fair	Good
	20	15	1.3	2,438	Fair	Good
Low-density ammonia dynamite, high velocity	65	50	1.2	3,353	Fair	Fair
	65	40	1.0	3,048	Fair	Fair
	65	30	0.9	2,896	Poor	Fair
	65	20	0.8	2,591	Poor	Fair
Low-density ammonia dynamite, low velocity	65	50	1.2	2,438	Fair	Fair
	65	40	1.0	2,286	Poor	Fair
	65	30	0.9	2,134	Poor	Fair
	65	20	0.8	1,981	Poor	Fair
Blasting gelatin	100	90	1.3	7,620	Excellent	Poor
Straight gelatin	90	80	1.3	7,010	Excellent	Poor
	60	60	1.4	6,096	Excellent	Good
	40	45	1.5	5,029	Excellent	Good
	20	30	1.7	3,353	Excellent	Good

Continued 2-1

Type	Weight strength, %	Bulk strength, %	Specific gravity	Detonation velocity, m/s	Water resistance	Fume class
Ammonia gelatin	80	70	1.3	6,096	Very good	Good
	60	60	1.4	5,334	Very good	Very good
	40	45	1.5	4,877	Very good	Very good
Semigelatin	65	60	1.3	3,658	Very good	Very good
	65	50	1.2	3,658	Very good	Very good
	65	40	1.1	3,505	Good	Very good
	65	30	0.9	3,200	Fair	Very good

2.3.2 Nitroglycerin

Nitroglycerin (NG), the first high explosive, is the sensitizer in dynamites and seldom is used alone, although it has been employed in a somewhat desensitized form for shooting oil wells. It has a specific gravity of 1.6 and a detonation velocity slightly over 7,620m/s. Its extreme sensitivity to shock, friction and heat make it a hazardous product to use.

2.3.3 Straight (Nitroglycerin) Dynamite

Straight (nitroglycerin) dynamite consists of nitroglycerin, sodium nitrate, an antacid, a carbonaceous fuel and, sometimes, sulfur. The term "straight" means that a dynamite contains no ammonium nitrate. The weight strength, which ranges from 20%—60%, indicates the approximate percentage of nitroglycerin or other explosive oil. The use of straight dynamite is limited because of its high cost and sensitivity to shock and friction. Fifty percent straight dynamite is referred to as ditching dynamite.

2.3.4 Ammonia Dynamite

Ammonia dynamite, also called extra dynamite, is the most widely used cartridged high explosive. It is like straight dynamite except that ammonium nitrate replaces part of the nitroglycerin and sodium nitrate. Ammonia dynamite is manufactured in grades of 20%—60% weight strength, although these grades are not truly equivalent to straight dynamites of the same weight strength (see properties in Table 2-1). Ammonia dynamite is slightly less sensitive to shock and friction. It is most commonly used in small quarries, in underground mines and as an agricultural explosive.

2.3.5 Low-Density Ammonia Dynamite

Low-density ammonia dynamite is manufactured in a weight strength of about 65%. The cartridge (bulk) strength ranges from 20%—50%, depending on the bulk density of the ingredients. A high-velocity series and a low-velocity series are manufactured. Low-density ammonia dynamite is most useful in very soft or prefractured rock, or where a coarse material, such as riprap, is required.

2. 3. 6 Blasting Gelatin

Blasting gelatin is a tough rubber-textured composition made by adding nitrocellulose, also called guncotton, to nitroglycerin. An antacid provides storage stability and wood meal improves sensitivity. Blasting gelatin emits large volumes of noxious fumes upon detonation and is very expensive. Sometimes called oil-well explosive, it is useful in deep wells where high heads of water are encountered. It is the most powerful of all nitroglycerin-based explosives.

2. 3. 7 Straight Gelatin

Straight gelatin is like blasting gelatin with sodium nitrate, carbonaceous fuel and, sometimes, sulfur added. It is manufactured in grades ranging from 90%—20% weight strength. Straight gelatin has been used mainly in seismic or deep-well work, where a lack of confinement or a high head of water may affect its velocity. To overcome these conditions, a high-velocity gelatin has been developed which is like straight gelatin except that it detonates near its rated velocity despite high heads of water.

2. 3. 8 Ammonia Gelatin

Ammonia gelatin, also called special gelatin or gelatin extra, is a straight gelatin in which ammonium nitrate has replaced part of the nitroglycerin and sodium nitrate. Weight strengths range from 80%—30%. Ammonia gelatin is suitable for underground work and as a toe load, primarily in small-diameter boreholes. The higher grades (70% or higher) are useful as primers for blasting agents.

2. 3. 9 Semigelatin

Semigelatin has a weight strength near 65%. The cartridge (bulk) strength ranges from 30%—60% with variations in the bulk density of the ingredients. Semigelatin is very versatile for small-diameter work where some water resistance is required. It is useful underground, and its soft plastic consistency makes it ideal for loading into holes drilled upward.

2. 3. 10 Permissible Explosives

Permissible explosives are designed for use in underground coal mines where the presence of explosive gases presents an abnormal blasting hazard. The Bureau of Mines has published requirements for determination of permissibility. Briefly stated, the specification of a permissible explosive are:

(1) The chemical composition furnished by the applicant must agree, within tolerances, with that determined by the Bureau.

(2) The explosive must pass a series of propagation tests.

(3) The air-gap sensitivity of 32mm cartridges must be at least 76mm.

(4) The explosives must pass nonignition tests when fired unstemmed into a mixture of natu-

ral gas, air and bituminous coal dust.

(5) The explosive must pass tests for nonignition when fired in a gallery of air and natural gas.

(6) The volume of poisonous gases produced by 1kg of explosive must not exceed 0.16 cu. m.

(7) The explosive must exhibit insensitivity in the pendulum friction test.

Most permissibles are nitroglycerin-based dynamites with sodium chloride or other flame depressants added. They may be granular or gelatinous and need not be nitroglycerin-based. The Bureau of Mines periodically publishes an up-to-date list of brand names and properties of permissible explosives.

2.3.11 Pneumatic Breaking Systems

Pneumatic breaking systems employing compressed air or liquid carbon dioxide are not really explosive in nature but are noted here because their primary application in mining has been in underground coal mines where restrictive explosive laws give such units an advantage over permissible explosives. In the compressed-air system, a high-pressure compress or feeds air to a receiver for storage. From the receiver, the air is piped to an air cartridge placed in a borehole. When the prescribed pressure is attained in the cartridge, a shear disk is ruptured, which causes the air to be discharged from a series of ports at the end of the cartridge. With the carbon dioxide system, liquid CO_2 is charged into a cartridge placed in a borehole. Electric current energizes a heating element that vaporizes the carbon dioxide. A mild-steel disk is sheared, and the gaseous carbon dioxide is discharged through a series of ports at the end of the cartridge, as in the compressed-air system. Both systems create a slow pushing action that breaks out the coal in coarse chunks. Pneumatic systems work well only in very weak formations. Use in a competent rock or with too large a burden may cause the cartridge to be expelled from the borehole at high velocity. The time-consuming procedure and expensive nature of pneumatic systems restricts their use to situations where blasting laws dictate or where a large lump product is the overriding consideration.

2.3.12 Liquid Oxygen Explosive (LOX) and Black Powder

Liquid oxygen explosive (LOX) and black powder merit mention because of their past importance. LOX consists of a cartridge of lampblack, carbon black or charcoal dipped into liquid oxygen just before loading. It derives its energy from the reaction of the carbon and oxygen to form carbon dioxide. LOX is fired with an ordinary detonator and attains velocities of 3,658 to 5,791m/s. The domestic consumption of LOX, primarily in strip coal mining, has been replaced by blasting agents, although LOX is used in foreign countries. Black powder, a mixture of potassium or sodium nitrate, charcoal and sulfur, dates from ancient times. Once the principal commercial explosive, black powder is extremely prone to accidental initiation by flame or spark. When initiated, it undergoes deflagration (rapid burning) at a very low velocity rather than detonating. It has a specific gravity of 1.6 or less, depending on granulation, has poor water resistance, and has a very poor fume rating. Black powder enjoys limited use in blasting dimension stone and slate where a minimum of shattering action is desirable. It is no longer manufactured in the United States.

2.3.13 Dry Blasting Agents

A blasting agent is a material or mixture that (1) consists of a fuel and an oxidizer, (2) is intended for blasting, (3) is not otherwise classified as an explosive and in which none of the ingredients is classified as an explosive, and (4) provided that the finished product, as mixed and packaged for use or shipment, cannot be detonated by means of a No. 8 blasting cap. The term nitrocarbonitrate is used synonymously with blasting agent and is a classification given under U. S. Dept. of Transportation regulations for packaging and shipping blasting agents. Fig. 2-8 shows the formulations of common dry blasting agents and slurries.

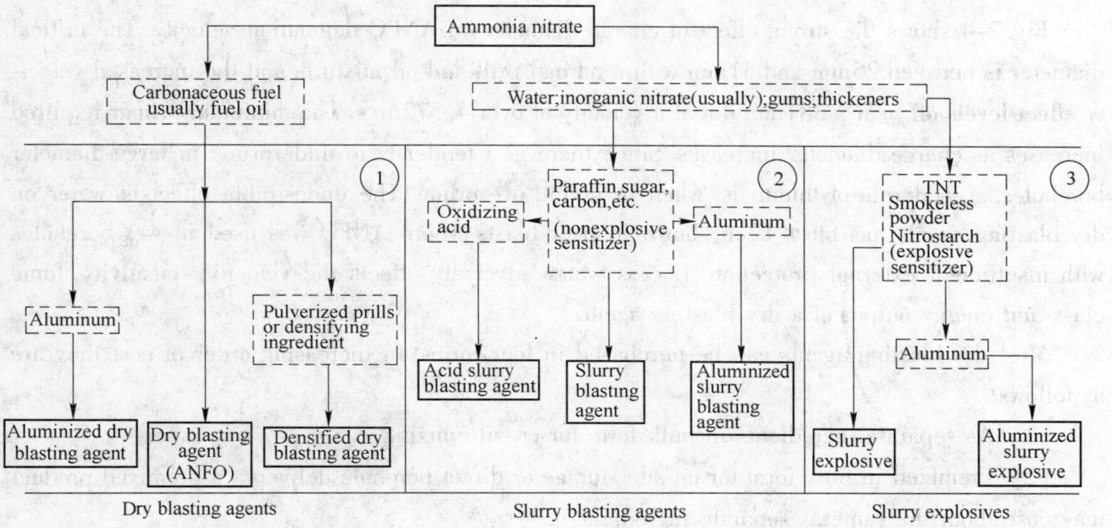

Fig. 2-8 Formulations of blasting agents and slurries

Early dry blasting agents employed solid carbon fuels combined with ammonium nitrate in various forms. Through experimentation it was found that diesel fuel oil mixed with a porous ammonium nitrate gave the most desirable results. Hence, the term ANFO (ammonium nitrate-fuel oil) has become synonymous with dry blasting agent. An oxygen-balanced ANFO is the cheapest source of explosive energy available today. A price of $0.08 to $0.11 per kg is common. Adding aluminum flakes to dry blasting agents increases the energy output but at an increase in cost. Aluminized dry mixes have application as primers for ANFO, as a high-energy toe load, and as the main-column charge where blasting is difficult.

It is difficult to give precise numerical values for the properties of dry blasting agents because they vary with ingredient particle size, density, confinement, charge diameter, water conditions, coupling ratio and size of primer, Yancik has prepared an excellent manual on this subject. Coupling ration is the percentage of the borehole diameter filled with explosive. Poured bulk products are completely coupled.

Cartridged products are partially decoupled and thus lose some efficiency. ANFO' s theoretical energy is optimized at oxygen balance (approximately 94.5% AN and 5.5% FO), where the detonation velocity is near 4,267m/s. Excess fuel oil can seriously reduce sensitivity to initiation. Specific gravities range from 0.5 to 1.15; 0.75 to 0.95 is the most common range. The lighter products are useful in easily fragmented rock or to eliminate the need for alternate decks of explosive and stemming where a low powder factor is desirable. The densified dry mixes are used as secondary primers or high-energy boosters, as a toe load, or as a full-column charge where blasting conditions are difficult. To obtain a bulk specific gravity much over 1.0, however, the blasting agent must be vibrated or otherwise compressed and loaded into cartridges or bags. Good coupling and confinement improve blasting-agent performance. The sensitivity of ANFO decreases with increased density. The "dead press" limit, above which detonation will not occur, is about 1.25.

Fig. 2-4 shows the strong effect of charge diameter on ANFO detonation velocity. The critical diameter is between 25mm and 51mm with a normal prill and oil mixture, and the increased velocity effect levels off near a 381mm dia at a velocity of over 4,572m/s. The minimum primer required increases as charge diameter increases. Since there is a tendency to underprime in large-diameter boreholes, a good rule-of-thumb is, when in doubt, overprime. The undesirable effect of water on dry blasting agents has often been shown in poor blasts where ANFO was used in wet boreholes with insufficient external protection. Excess water adversely affects the velocity, sensitivity, fume class and energy output of a dry blasting agent.

Most dry blasting agents can be purchased in four forms. In increasing order of cost they are as follows:

(1) As separate ingredients in bulk form for on-site mixing.

(2) Premixed in bulk form for on-site storage or direct borehole delivery (a premixed product may cost about the same as separate ingredients).

(3) In paper, polyethylene or burlap packages.

(4) In rigid cartridges or metal cans.

Rigid cartridges and cans are the most expensive forms and eliminate the advantage of direct borehole coupling. These should be used only where borehole conditions dictate. Because improper ingredient proportions or an insufficiently mixed product cause inefficient detonation and poor fume qualities, thorough mixing and close quality control should be the rules in an on-site mixing operation. The use of a colored dye in the fuel gives a visual check on mixing and also makes the blasting agent more easily visible in case of a misfire.

"Dry slurries", which are essentially dry blasting agents that have been coated with a water barrier or a material that forms a seal when exposed to water, are under study. They show promise of being useful in moderately wet conditions. Experimental products that have increased density due to a larger proportion of liquid ingredients also have been studied.

2.3.14 Slurries (Water Gels)

The terms "slurry" and "water gel" have been used interchangeably, although some writers reserve the term "water gel" for products that have been cross-linked to a semisolid jellylike consistency and use "slurry" for more fluid products. Here, the terms are considered synonymous. A slurry is a mixture of oxidizers, such as sodium nitrate and ammonium nitrate, a fuel-sensitizer, either explosive or nonexplosive, and water (usually about 15%). This mixture is thickened with a guar gum and cross-linked to set up an impenetrable barrier to prevent leaching of the soluble oxidizing salts by water in the borehole. As a result, most slurries have excellent water resistance. The variety of possible slurry formulations is almost infinite. The list of possible fuel-sensitizers is especially long, although carbonaceous fuels, aluminum, trinitrotoluene (TNT) and smokeless powder have been the most common. Slurries containing explosive sensitizers, such as TNT, have been incorrectly referred to as blasting agents contrary to the National Fire Protection Association's definition. A slurry explosive is one which is sensitive to a No. 8 blasting cap or which contains an ingredient classified as an explosive. A slurry blasting agent must satisfy the definition in the "Dry Blasting Agents" section. Actually, some explosive slurries are less sensitive than fine-grained dry blasting agents or slurry blasting agents containing pigmented-grade aluminum. In this discussion, slurry blasting agents and slurry explosives are treated collectively.

Slurries commonly are identified by their fuel-sensitizer, such as aluminized slurry, TNT slurry or smokeless-powder slurry. Rating the strength of a slurry by its content of explosive, aluminum or other sensitizer can be misleading. Excess sensitizer causes an oxygen deficiency in the mixture which, in effect, wastes part of the fuel. Except for their water resistance and higher densities and bulk strength, slurries are similar in most ways to dry blasting agents. Good oxygen balance, decreased particle size and increased density, increased charge diameter, good confinement and coupling, and adequate priming all increase their efficiency. Although slurry blasting agents tend to lose sensitivity near a specific gravity of 1.35, explosive (TNT) slurries function well at a density of 1.6. The effect of charge diameter on slurries (Fig. 2-4) is not as pronounced as on ANFO.

Slurries can be delivered as separate ingredients for on-site mixing, premixed for bulk loading, or cartridged. Slurries for extruding into small-diameter boreholes are a recent development. A slurry can be cap-sensitive with a critical diameter below 12.7mm or it can be very insensitive, requiring a larger charge diameter and a powerful primer to function efficiently. The consistency of a slurry can be anywhere from a liquid to a cohesive gel. Explosive slurries are dense and develop high detonation pressures, which make them excellent for blasting hard rock. The future of explosive slurries is in doubt, however, because TNT is no longer manufactured commercially in the United States, and smokeless-powder slurry production depends on the availability of Government-surplus smokeless powder. Metallized (aluminized) slurries develop more energy than other commercial slurries and perform well in hard dense rock. However, they are somewhat expensive, particularly those with high metallic contents. Slurries containing neither aluminum nor explosive sensitizers are the cheapest, but they are also the least dense and powerful. In wet conditions where dewatering is not practical, low-cost slurries offer competition to ANFO. Some operators have successfully mixed dry blasting agents and slurries and retained sufficient water resistance for moderate water conditions, although this technique should be used cautiously.

2.4 Blasting Accessories

2.4.1 Blasting Caps

The electric blasting cap is the most commonly used device for initiating high explosives. It may be inserted directly into the explosive cartridge or cast primer, or be used with detonating cord. The advantages of electric blasting caps are safety in handling, availability of fairly accurate delay periods, and ability to closely control the time of firing. In populated areas, noise may be reduced by initiating the charge in the borehole rather than using a detonating cord trunkline. Stray currents, such as those caused by power cables lying on the ground, should be avoided. Lightning also is a hazard. Radio-frequency energy is a potential hazard, although somewhat remote. Because current requirements vary from brand to brand, mixing of brands within a circuit can cause a misfire and is not recommended. Two series of delay electric blasting caps are available: short or millisecond delays, with delay intervals of 25 to 50 m-sec, and slow or half-second delays. In normal blasting, millisecond delays yield good breakage and reduce ground vibrations. Slow delays are used primarily underground and in tunnel work, where they provide sufficient time for rock movement between delay periods. A delay series for use in underground coal mines also is available.

The cap and fuse is an alternate means of initiating explosives. Safety fuse, consisting of a core of potassium-nitrate black powder enclosed in a covering of textile and waterproofing compound, is inserted into the open end of the cap, butted to the explosive charge, and crimped to form a tight bond. Most varieties of fuse burn at 131 sec per m; some burn at 98. Cap and fuse is used underground where rotational firing is necessary and can provide an unlimited number of delay periods. The combination also is useful in single-shot work, as in agriculture and secondary blasting. Fuse caps require cautious handling because the highly sensitive explosive is exposed at the open end of the cap. Mishandling of the fuse, a change in altitude, or a change in confinement can change the burning rate. The cap-and-fuse combination, however, eliminates the hazard of stray electricity and the need for a power source and associated lead wires.

2.4.2 Detonating Cord

Detonating cord, or detonating fuse, consists of a core of high explosive, usually pentaerythrite tetranitrate (PETN), contained in a waterproof plastic sheath and enclosed in a reinforcing covering of various combinations of textile, plastic, wire and waterproofing. Core loads of 3 to 1,312 grains per m are available. All grades are initiated with a blasting cap and have a detonation velocity near 6,401m/s. Originally, a core load of 164 grains per m of PETN was most common, but the 82 grade has been found adequate for most uses. Some manufacturers recommend the 82 grade for use with relatively insensitive cast primers. Its insensitivity to shock and friction makes detonating cord ideal for use as a downline and trunkline in primary blasting. Since the cap need not be put into the circuit until immediately before blasting, most of the hazard of premature detonation is eliminated. Eighty-two or 164-grains-per-m detonating cord will not detonate blasting agents under normal conditions. Detonating fuse will propagate when wet but should be initiated at a dry end. Millisecond delay connectors are available for use where delay blasting is required. They are connected between two ends of detonating cord in a trunkline and give an unlimited number of delay periods.

In some blasting, nonelectric bottom initiation is required where normal detonating cord would initiate a main charge of high explosive at the top or would cause deterioration in a blasting agent. To alleviate this, an assembly with a blasting cap attached to low-energy (9 to 16 grains-per-m) detonating cord is available. This assembly also is useful where stray currents are a hazard, as in salt or other conductive rock, or when loading pneumatically with bottom initiation. Low-energy detonating cord, with special connectors, also can be used as a trunkline where noise is a problem.

2.4.3 Primers and Boosters

A primer is a charge of explosive placed within the main charge of blasting agent or insensitive explosive to initiate detonation. By original definition, the primer included the initiating device, such as a blasting cap or detonating cord. A booster, a charge of explosive placed at intervals in the borehole to reinforce the detonation of a column of blasting agent, differed from a primer in that it did not contain an initiating device. However, new terminology suggested by the Institute of Makers of Explosives and included in the Bureau of Mines Health and Safety regulations refers to a charge containing an initiating device as a "capped primer" and uses the terms "primer" and "booster" synonymously to describe a charge that does not contain an initiating device. Although various products have been used as primers and boosters, an explosive with a high detonation pressure, such as a high-strength ammonia gelatin or a cast military explosive (Composition B or pentolite), is recommended. Military explosives are described in a Department of the Army publication. Although small 0.4536kg cast primers are popular, even in large boreholes, a primer functions best when its diameter is near that of the borehole. A two-stage primer, with a charge of high-energy dry blasting agent or slurry poured around a cast primer or ammonia gelatin, is a very efficient unit. In Sweden, in small-diameter work, excellent results are reported with a high-strength blasting cap used to initiate ANFO, and thus eliminate the need for a primer. The blasting cap reportedly contains about 1 g of high explosive. This system has a safety advantage, although blasting efficiency can be reduced through inadequate priming.

2.5 Explosive Prices

The prices of explosive products are difficult to pinpoint. They vary with time, economic conditions, type and grade of product, distance from the supplier, competition in a given area, type of packaging and volume of purchase. Fig. 2-9 gives approximate price ranges for various products as of 1971. More accurate prices are available from suppliers upon request.

2.6 Choosing an Explosive

There are no charts to which a consumer can refer to choose the proper explosive or blasting agent. This choice is dictated by field conditions and, in the final analysis, economic considera-

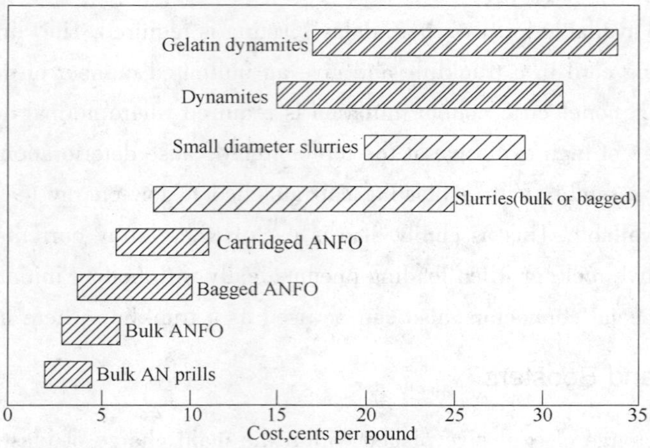

Fig. 2-9 Generalized prices of explosives and blasting agents

Note: 1lb = 0.4536kg

tions. When cartridged nitroglycerin explosives were predominant, impedance matching between rock and explosive was a popular theory. The impedance of a rock is the product of its density and sonic velocity. The impedance of an explosive is the product of its density and its detonation velocity. Thus a dense competent rock called for a dense high-velocity explosive.

The advent of dry blasting agents and slurries has changed the economic picture. ANFO at \$0.08 to \$0.11 per kg is difficult to complete with, despite its low impedance. Slurries, although more expensive than ANFO, offer better economy than cartridged nitroglycerin explosives for most large operations. Added to these lower product costs is the economy of bulk loading techniques.

For most operators, then, the choice becomes one between dry blasting agents and slurries. Authorities differ on the relative merits of the two. Because of low product cost and efficient bulk-loading techniques, however, the overwhelming majority of operators favor dry blasting agents (Fig. 2-5). ANFO's disadvantages of (1) low density, (2) lack of water resistance and (3) inefficient detonation in small diameters, can be overcome by (1) using a larger borehole, (2) dewatering and using a plastic borehole liner (not recommended underground) and (3) high-velocity pneumatic loading.

In extremely difficult blasting conditions or where drilling is expensive, economics may dictate the use of a slurry. Some operators compromise and use a slurry or densified high-energy dry mix in the lower portion of the borehole and ANFO as a top load.

Controlled in-the-mine research with careful cost analysis is the best way to determine the optimum explosive product to use. The choice of explosive will affect costs in the drilling, secondary breaking, loading, hauling and crushing subsystems, as well as the blasting subsystem. Although a detailed cost analysis will favor ANFO in most cases, there are situations where the extra cost of a slurry is more than returned through savings in the other operational subsystems.

2.7 Blasthole Loading Methods

The method used to load a borehole is determined primarily by the borehole diameter and the type of explosive used. For this discussion, boreholes are categorized as small-diameter (less than 102mm), which may be drilled either horizontally or vertically, and large-diameter (102mm or more), which normally are drilled vertically. Where a sufficient volume of explosive is used, the operator may find it advantageous to contract loading and firing to an explosives supplier or a blasting contractor. In some highly competitive areas, large customers receive this service free, although this practice has become less prevalent.

2.7.1 Loading Small-Diameter Boreholes

In loading small-diameter boreholes, the primer should be placed at the bottom of the hole for maximum confinement of detonation and also to guard against leaving undetonated explosive in the borehole if it should become plugged during loading. To minimize violence (airblast and flyrock), the distance from the top of the explosive column to the borehole collar should be 14 to 28 times the borehole diameter. Where flyrock and airblast are not problems, and low-cost ANFO is used, boreholes sometimes are loaded nearly to the collar, although the last 0.3m or 0.6m of blasting agent is practically wasted. The use of stemming reduces airblast and helps somewhat to confine the explosive gases. In vertical boreholes, stemming with drill cuttings is relatively easy. In horizontal boreholes, however, the operator must weigh the advantages of stemming against the time and cost involved. In underground coal mines, permissible standards require that each borehole be stemmed with a noncombustible material. Water-filled stemming cartridges are useful underground in horizontal boreholes in reducing dust from blasting and in preventing ignition of atmospheric gases. Deck loading in small-diameter boreholes is of doubtful benefit since the calculated distance from the top of the main charge to the collar usually is small.

2.7.1.1 Nitroglycerin-Based Explosives

Nitroglycerin-based explosives usually are loaded by hand and tamped one or two cartridges at a time to improve coupling. Although pneumatic loading of cartridges through a tube has been practiced in Sweden, it has not found acceptance in this country. Bulk pneumatic loading of high explosives is a dangerous practice. When loading cartridges, using the largest diameter cartridge compatible with the borehole diameter will improve the explosive's coupling and hence its efficiency. Because of the high price of nitroglycerin-based explosives, and the high cost of loading by the cartridge, use of these products has declined.

2.7.1.2 Dry Blasting Agents

Dry blasting agents may be pneumatically loaded, poured from the bag, or loaded by the cartridge into small-diameter boreholes. Poured charges in diameters less than 102mm lose some efficiency because of the diameter effect on velocity, although the benefits of good coupling are achieved. Pneumatic loading improves ANFO performance by pulverizing the prills and giving higher loading densities and greater sensitivity. High-and low-Pressure pneumatic systems are available. The low-pressure type propels the prills into the borehole at a low velocity and a high volume rate, loading ANFO at its poured density or slightly above with a minimum of prill breakage. The Venturi, or high-pressure system, propels the ANFO at a high velocity and low volume rate, pulverizing the prills and giving bulk loading densities near 1.00. The benefits of high-velocity pneumatic loading increase as the borehole diameter approaches 25.4mm(Fig. 2-10).

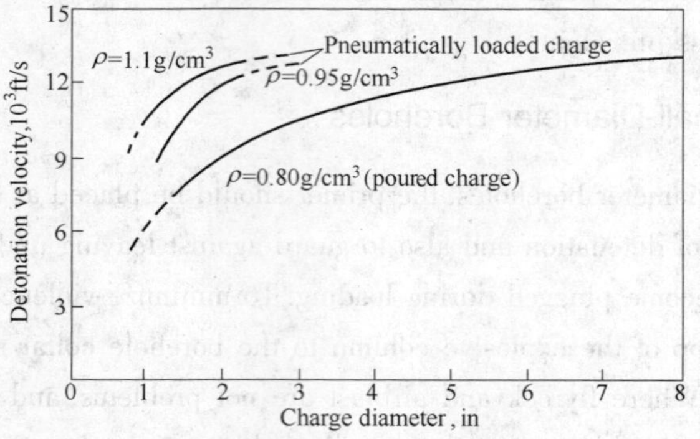

Fig. 2-10 The effect of pneumatic loading on ANFO in small-diameter boreholes (typical values)

Note: 1ft/s = 0.3048 m/s; 1 in = 25.4mm

The pneumatic loading tube is useful for blowing standing drill water from a horizontal borehole. However, if the borehole is "making water", external protection for the ANFO is required. Loading inside a plastic borehole sleeve is not recommended for underground work because of the static electricity hazard and toxic fumes generated. If plastic-sleeve protection with pneumatic loading in well-ventilated locations is required, a nonelectric detonating system should be used because the insulating effect of the sleeve is likely to cause a buildup of static electricity. Low-energy detonating cord systems are recommended here because standard 82-and 164-grain detonating cord can have a degrading effect on dry blasting agents in a small-diameter borehole. Even without a plastic sleeve, pneumatic loading presents a potential static electricity hazard. Some state laws prohibit pneumatic loading over blasting-cap leg wires under any circumstances. Where this practice is permitted, the use of a specially constructed semiconductive loading tube and a properly grounded loading vessel will minimize the hazard of static electricity. Reactivity with pyrite-bearing rock is another potential hazard with ANFO.

Cartridged dry blasting agents are useful where water protection is required, but they are more expensive and the degree of coupling is reduced. In water, cartridges with a specific gravity less than 1.00 will float unless preventive measures care taken. A high-velocity primer that nearly fills the borehole diameter is recommended for all blasting agents. In long small-diameter boreholes, occasional boosters placed up the borehole are good insurance. The use of blasting agents is not permitted in underground coal mines. The Bureau of Mines has published safety recommendations for handling ammonium-nitrate blasting agents.

2. 7. 1. 3 Slurries

Slurries recently have been developed for small-diameter blasting. A finely flaked paint-grade aluminum gives a slurry with a small critical diameter. Semirigid gels packaged in polyethylene tubing with diameters as small as 25. 4mm are loaded in a manner similar to hand loading of cartridged nitroglycerin-based explosives. Some of these water gels are cap-sensitive and some are blasting agents, requiring priming similar to dry blasting agents. Machines for bulk extrusion of slurries into small-diameter boreholes are being developed. As these machines and the compatible extrudable slurries become more highly developed they will offer additional competition to nitroglycerin-based high explosives in the small-diameter field.

2. 7. 1. 4 Permissible Blasting Practices

In underground coal mines, the loading and firing methods, as well as the explosive type, must be permissible. This discussion is intended to point out some basic differences between ordinary and permissible blasting procedures, but the list is not all-inclusive. Persons involved in blasting coal underground should contact a U. S Bureau of Mines District Office of Coal Mine Health and Safety. Permissible regulations, some of which are still pending, include the following:

(1) Black powder or other nonpermissible explosives, including detonating cord, may not be stored or used underground.

(2) Unconfined shots may not be fired.

(3) A borehole deeper than the cut must be tamped with incombustible stemming material to the depth of the cut to insure that the explosive is not fired on the solid.

(4) Charge weights may not exceed 1. 36kg per borehole.

(5) Boreholes should have a minimum burden of 457mm in all directions. If this specification can not be met, the charge weight should be reduced to prevent underburdened shots.

(6) The minimum depth of cut should be 1. 07m.

(7) A round may not contain more than 20 shots unless special permission is obtained.

(8) Aluminum-cased detonators may not be used. Leg wires should not be more than 4. 88m long, or of equivalent resistance. Zero-delay detonators should not be used in a circuit with millisecond-delay detonators. Delays between successive shots should not exceed 75 m-sec, and no more than two holes per delay should be fired. The total time lapse between the first and last shot in a round should not exceed 500m-sec.

(9) All holes must be bottom primed with the cap at the back of the hole.

(10) Cartridges must be loaded in a continuous train, in contact with each other, and should not be deliberately crushed or deformed.

(11) All blastholes must be stemmed with incombustible material. Holes deeper than 1. 2m must contain at least 457mm of stemming and holes less than 1. 2m deep should be stemmed for at

least half their length. Water stemming bags should be at least 381mm long and should have a diameter within 6mm of the borehole diameter.

(12) Shots must be fired with a permissible unit of adequate capacity. Of course, all of the normal blasting safety rules apply.

2.7.2 Loading Large-Diameter Boreholes

Large-diameter boreholes (102mm or more) normally are drilled vertically, or nearly so. In a large-diameter borehole, the desired primer location may be anywhere from the bench toe to the center of the charge. Initiation can be by an electric blasting cap inserted into the primer or by detonating cord connected to the primer and attached to a trunkline. In a large-diameter hole, ordinary detonating cord will not adversely affect the properties of a blasting agent. With cap-sensitive explosives, initiation with a detonating cord downline will give top initiation with consequent premature loss of confinement of detonation gases. The distance from the top of the charge to the borehole collar should be 14 to 28 times the charge diameter. The larger values are required for top initiation and where flyrock is a problem.

To reduce airblast and confine the gases of detonation, stemming with drill cuttings is most common. The stemming normally is hand-shoveled, although with very large-diameter boreholes (near 381mm) a small bulldozer may be used. In some operations, specially sized stemming material is brought in, although it is doubtful whether this is economically justified under normal circumstances. Deck loading, where zones of explosives and stemming are alternated, was a common practice with expensive cartridged explosives to reduce costs. With cheaper lower-density blasting agents, this practice is not usually worthwhile. Occasionally, a small deck charge may be used near the borehole collar to help fracture a difficult cap rock.

2.7.2.1 Nitroglycerin-Based Explosives

Because of their high cost, nitroglycerin-based explosives seldom are used in large-diameter boreholes except as primers or boosters. Blasting agents do an equal or better job at a fraction of the cost. Large-diameter cartridges of high explosives usually are available only on special order.

2.7.2.2 Dry Blasting Agents

Dry blasting agents may be loaded from bulk trucks, poured from the bag or loaded by the cartridge. Pneumatic and auger loading systems are used. The blasting agent may be premixed or the prills and oil may be metered together at the borehole. Auger loading gives the highest loading rates. Side-boom augers are satisfactory for loading single rows. Overhead-boom augers are preferable for loading several holes from a single setup and for multiple-row shots. In rough terrain, where it is difficult to get a truck near the borehole, pneumatic units with long loading hoses are advantageous, although the loading rate is lower than with augers. Simple units in which the blasting agent is gravity-loaded from the truck body can be used where terrain is no problem. Where bulk-loading is not feasible, hand-pouring from 22.68kg bags is an alternative and gives the same complete coupling as bulk-loading. High-density dry products (specific gravity, 1.05 to 1.15) in cartridges or waterproof bags have been used in wet conditions. Here the advantage of direct borehole coupling is lost, and broken package can cause desensitization.

In wet conditions an increasingly popular procedure is to pump the water from the borehole, insert a plastic borehole liner and load a dry blasting agent inside the liner. Here, water protection is assured without losing the benefits of direct borehole coupling. Since plastic liners increase toxic fumes they should be used only where ventilation is adequate. In competent rock, boreholes may be dewatered by detonating a small (1 to 5kg) charge in the bottom of the hole, although the danger of borehole cave-ins exists when using this procedure. The borehole should be allowed to cool before the ANFO is loaded. The type and size of primer used in a large-diameter borehole is important. Although small cast primers or cartridges of gelatin have been used alone for priming, most authorities agree that a charge of high-energy dry product or slurry poured around the primer increases blasting efficiency. Boosters of high-energy product sometimes are spaced up the borehole to reinforce detonation. In diameters above 230mm, where blasting agents detonate efficiently, boosters probably are of little value.

2.7.2.3 Slurries

In large boreholes, slurries are loaded either in bulk or as a polyethylene-bagged product. Bulk loading gives slightly better coupling and better loading economy, although bag loading is satisfactory for smaller operations. Charging, stemming and priming requirements for most slurries are the same as for dry blasting agents. Bulk pump trucks may bring a plant-mixed slurry to the site or separate components for on-site mixing. On-site slurry mixing is more complex than ANFO mixing and should be done by a competent explosive distributor rather than the consumer. Plant mixing permits closer quality control in the blending of ingredients, whereas on-site mixing permits variation of the product from hole to hole or within a given borehole. The slurry is pumped as a liquid, and the cross-linking ingredient is added just as the slurry enters the loading hose. Cross-linking begins in the hose and is completed in the borehole. A hose diameter below the critical diameter of the slurry will prevent an accidental detonation from propagating back to the truck. Slurries are very insensitive, however, and the chance of an accidental detonation is remote.

2.8 Controlled Blasting Techniques

Controlled blasting is used to reduce overbreak and minimize fracturing of the rock at the boundary of an excavation. Controlled blasting is art rather than science since blast dimensions for a specific job usually are arrived at by trial and error. The four basic controlled blasting techniques are: line drilling, presplitting, cushion blasting and smooth blasting. Table 2-2 gives some average specifications for controlled blasting design. To design a controlled blast for a specific project, an explosive manufacturer or a blasting consultant should be employed, because geological variables may cause spacings to vary considerably from these averages.

Table 2-2 Average specifications for controlled blasting

	Hole diameter, mm	Spacing, m	Burden, m	Explosive charge, kg/m
Line drilling	51	0.1—0.2	—	—
	76	0.2—0.3	—	—
Presplitting	38—44	0.3—0.5	—	0.12—0.37
	51—64	0.5—0.6	—	0.12—0.37
	76—89	0.5—0.9	—	0.19—0.74
	102	0.6—1.2	—	0.37—1.12
Cushion blasting	51—64	0.9	1.2	0.12—0.37
	76—89	1.2	1.5	0.19—0.74
	102—114	1.5	1.8	0.37—1.12
	127—140	1.8	2.1	1.12—1.49
	152—165	2.1	2.7	1.49—2.23
Smooth blasting	38—44	0.6	0.9	0.18—0.37
	51	0.8	1.1	0.18—0.37

2.8.1 Line Drilling

Line drilling, the earliest controlled blasting technique, involves drilling a row of closely spaced holes along the final excavation line, providing a plane of weakness to which to break. Line drill holes, 50 or 70mm in dia, usually are spaced 2 to 4 diameters apart and contain no explosive. The blastholes adjacent to the line drill holes normally are loaded lighter and are on closer spacing than the other blastholes. The maximum depth for line drilling is about 9m. Drill hole deviation is the limiting factor. As with all controlled blasting techniques, line drilling can be done in a series of lifts with a sufficient offset between benches to permit drilling for the next lower bench. Line drilling involves no blasting in the final row of holes, and thus minimizes damage to the final wall. The chief disadvantage is the high drilling cost.

2.8.2 Presplitting

Presplitting, sometimes called preshearing, involves a single row of boreholes, usually 50 to 100mm in dia, drilled along the final excavation at a spacing of 6 to 12 borehole diameters. Dynamite cartridges 25 to 38mm in size on 300 to 600mm centers usually are string-loaded on detonating cord, although special small-diameter cartridges with special couplers are available for total column loading. In unconsolidated formations, closer spacings with lighter powder loads are required. The bottom 600 to 900mm borehole usually is loaded somewhat heavier than the remainder. Stemming between and around the individual charges is optional. The top 600 to 900mm of borehole is not loaded, but is stemmed. The depth that can be presplit is limited by hole alignment, with 15m being about maximum. The presplit holes are fired before the adjacent primary holes to provide a fracture plane to which the primary blast can break. In presplitting, it is difficult to determine the results until the adjacent primary blast is shot. For this reason, presplitting too far in advance is not recommended. Several states have developed standards for presplit blasting in construction work. Presplitting seldom is done underground.

2. 8. 3 Cushion Blasting

Cushion blasting involves drilling a row of 50 to 150mm-dia boreholes along the final excavation line, loading with a light well-distributed charge, completely stemmed, and firing after the main excavation is removed rather than before, as in presplitting. The burden on the holes is slightly larger than the spacing. Wedges may be used to abut the charges to the excavation side of the borehole and minimize damage to the final wall. Explosive loading is similar to that in presplitting. Cushion blasting has been done to depths near 30m in a single lift with the larger-diameter boreholes because alignment is more easily retained. Cushion blasting seldom is done underground.

2. 8. 4 Smooth Blasting

Smooth blasting is the underground counterpart of cushion blasting. At the perimeter of the tunnel or drift, closely spaced holes with a burden-to-spacing ratio near 1.5:1 are loaded with light well-distributed charges. Smooth blasting differs from cushion blasting in that (1) except at the collar, the charges are not stemmed and (2) the perimeter holes are fired on the last delay in the same round as the primary blast. Total column loading is most common, although spacers may be used. The back of the borehole is loaded somewhat more heavily. The holes are stemmed to prevent the charges from being pulled out by the detonation of the previous delayed holes. Smooth blasting reduces overbreak in a drift and also provides a more competent back requiring less support. It involves more perimeter holes than does normal blasting.

Combinations of controlled blasting techniques are used. In unconsolidated rock, line drilling sometimes is desirable between presplit or cushion blastholes. Corners sometimes are presplit when cushion blasting is used.

Professional Terms

explosive 炸药
nitroglycerine explosive 硝化甘油炸药
ammonium nitrate explosive 硝铵炸药
gelatin 胶质炸药
slurry 浆状炸药
ANFO (ammonium nitrate-fuel oil) mixture 铵油炸药
strength 威力(猛度)
detonation velocity 爆速
density 密度
detonation pressure 爆炸压力
borehole pressure 炮孔压力
water resistance 抗水性
fume class 炮烟级别(按炮烟量分级)
sensitivity 感度
blast 爆破
blasting accessory 爆破器材
blasting cap 雷管
detonating cord 导爆索
primer 起爆药包

booster 传爆药包
controlled blasting 控制爆破
presplitting 预裂爆破
cushion blasting 缓冲爆破
smooth blasting 光面爆破

Exercises

2-1 请举出四种常用的炸药。
2-2 简述炸药的性质。
2-3 常用的爆破器材有哪些?
2-4 简述控制爆破原理。

3 Haulage of Non-ferrous Metal Mines

本章介绍有色金属矿山的地下运输和地面运输。地下运输包括矿石运输、材料运输和人员运输，涉及矿车、轨道、机车、卡车和地下自卸设备等。地面运输包括铁路运输和汽车运输。

3.1 Underground Haulage

One of the most important considerations in the efficient operation of an underground mine is the haulage system. Often the determining factor between profit or loss is the quick removal of ore and waste from the working places to secondary and main-line haulage areas, and so on to the outside. Important, too, is moving supplies from the surface to the working faces so that the loading process can continue with little interruption. Men must be transported in a rapid but safe manner.

It has been through the efficient use and generally the combination of mine cars, track, belt conveyors and rubber-tired haulage equipment that underground operations have been able to compete with the more popular strip or surface mining. As near-surface ore bodies are exhausted, underground haulage will play an expanding part in economically furnishing the world' d needs for all types of minerals.

The choice of underground haulage equipment, wherever possible, should be one that will give the smallest overall cost of ore removal during the life of the mine while meeting necessary safety requirements.

3.1.1 Mine Cars and Track

Many mines today use a combination of belt and rail haulage. Even if a belt system is used to carry the product from the face to the surface, track is used to transport workmen and supplies.

A track system, when properly installed, will provide interruption-free and safe haulage.

Schrecengost lists the following as major advantages of track haulage:

(1) Safety in transporting men in and out of the mine in personnel cars.

(2) Easy and rapid transportation of supervisory personnel.

(3) A temporary shutdown occasioned by a roof fall or power failure along the haulage system will not shut down the production areas.

(4) Quick availability of repair parts and supplies.

(5) Large pieces of coal or rock can be handled without damage to the haulage equipment.

(6) In areas where quality fluctuates noticeably, different cars may be separated out for special preparation.

(7) Rock or equipment may be loaded and moved out of the mine without interference with the production or preparation of the coal.

3.1.1.1 Mine Cars

Mine cars with steel bodies are used in all types of present-day mining. Wooden cars usually are more bulky and less resistant to wear and damage, but are more easily repaired. Regardless of the type of mine car selected, it is most practical to standardize on one type or make to simplify repairs and limit the amount of spare parts necessary to stock.

Rigid-body flat-bottomed cars are simpler and usually lower than others of equal capacity. Advantages are: ease of loading because of the low sides, simplicity, cheapness and high ratio of capacity to weight. The disadvantages include the fact that they can be dumped only at a tippler or on track dumps at fixed places.

Fig. 3-1 Rigid car

Fig. 3-1 shows an all-steel rigid car popular in coal mining. It comes in sizes to 101.6 tonnes, and is emptied by a rotary dump. Bottom-dump cars (Fig. 3-2) are used in many mines because of dumping simplicity. The side-dump Granby car shown in Fig. 3-3 is a heavily reinforced car used in taking large pieces of ore from chutes.

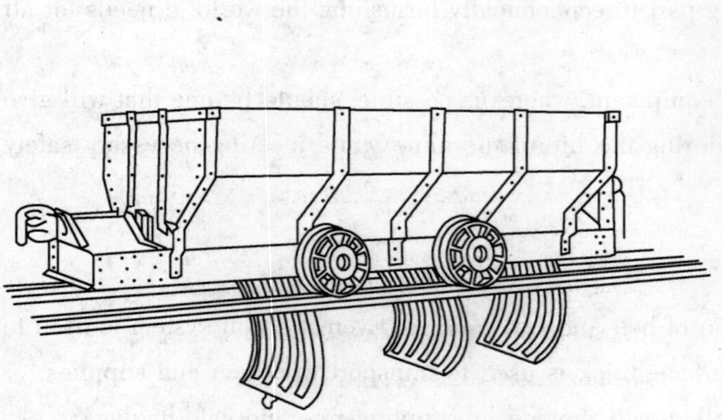

Fig. 3-2 Automatic drop-bottom car

Fig. 3-3 Granby car in dumping position

3.1.1.2 Mine Track

Steel rails of ASCE design usually are employed in most underground mines. The minimum weight that should be used is controlled by the locomotive weight. A simple rule to determine rail weight in kilograms per meter is 5kg per m for each metric ton of weight on a wheel. Suggested rail weights for different-sized locomotives are shown in Table 3-1.

Table 3-1 Suggested rail weights for mine locomotives

Locomotive weight, t	Weight of rail per meter, kg/m		
	4-wheel locomotive	6-wheel locomotive	8-wheel locomotive
2	10		
4	12		
6	15		
8	15		
10	20	15	
13	30	20	
15	30	20	
20	30	30	
27	40	35	30
38	50	42	35
53		50	45

Steel rails are made in standard series. For mine track, the best should have a thick web and a broad base because of the use of light ties and the frequency of corrosive water. Table 3-2 shows rail sections that most nearly meet these requirements.

Table 3-2 Rail sections

Weight, kg/m	$b = d$①, mm	c①, mm	t①, mm	Size of hole, mm	Track bolt size, mm	Weight splice bars, per pair, kg
10	67	34	6	19	13×51	2
12	70	38	7	21	16×57	3
15	79	43	8	21	16×64	4
17	84	44	9	21	16×70	5
20	89	48	10	22	16×76	7
22	94	51	11	22	19×76	9
25	98	54	11	25	19×83	12
27	103	57	12	25	19×89	13
30	108	60	12	25	19×89	15
32	113	61	13	25	19×95	16
35	117	62	13	25	19×95	25
40	127	64	14	29	19×105	29
42	132	65	14	29	19×108	31
45	137	67	14	32	22×108	34
50	146	197	14	32	25×108	40

Note: Rails over 25kg per m are 10m long with 10% shorts; 25kg or less, 9 m long with 10% shorts. Splice bars for rails 32kg and over have six holes; less than 32kg, four.

① Letters are items in Fig. 3-4.

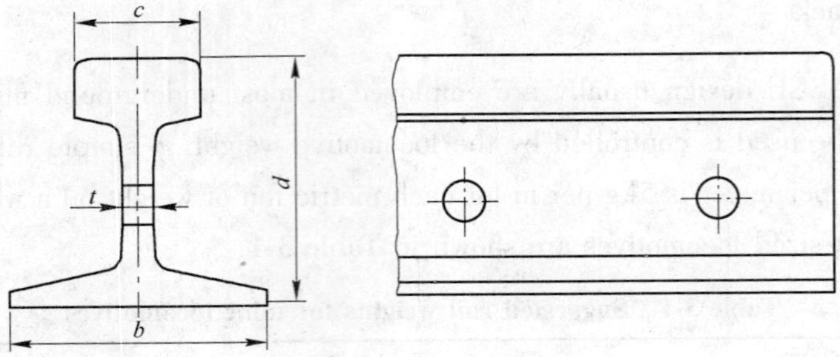

Fig. 3-4 Rail section

Spacing of holes in rails and splice bars varies with different makers for rails below 18kg, and should be specified when ordering. It is suggested that manufacturers' catalogs be consulted before detailing.

Rail joints should have as nearly as possible the same strength and stiffness as the solid rail. Expansion is allowed for by leaving spaces between rail ends, using splice bars with slotted bolt holes, and drilling the rail holes 3 to 6mm larger than the bolts. Flat splice bars are used for rail up to 18kg; for rails over 18kg, the angle-bar type is better.

Wood ties for most mine work should have a length twice the track gage, be at least 6mm thicker than the spike length, and 1.375 times spike length in width. Steel ties have the advantage of lightness, strength and durability. They require less headroom and are not a fire hazard.

Spacing of ties underground will range from 0.4m on some main haulageways with soft beds to 1.2 and even 1.8m centers in rooms or stopes. An excessive span with light rails or heavy wheel loads greatly increases resistance, and bending of the rail loosens spikes and cuts ties. Track stringers may be necessary where road beds are soft. Life of ties usually depends on the number of times relaying is necessary. Preservative treatment is economical only in main entries or long adits.

Track ballast may be broken stone, gravel, tailings, waste, cinders or slag. Special ballast seldom is needed for mine track. The floor usually is hard, and care of the drainage ditches usually will remedy soft spots. Ballast will not help swelling or heaving floor. Track gage should not be less than half of the extreme width of car or locomotive. A crude empirical rule, where wheel base does not greatly exceed gage is: gage < $\sqrt{R/4}$, where R is the shortest radius.

1. Bonding

The contact resistance of ordinary rail joints is so great as to be equivalent to an open circuit. The best bonds should be used under splice bars, and rails should be cross-bonded about every third length, the rail circuit being connected to the negative pole to minimize corrosion. Electric welding is the best modern method of bonding.

2. Laying Out Curves, Switches and Crossings

In laying track around curves, the rails should not be forced into position. They should be bent by a rail bender to conform to the radius of the curve. The best procedure is to buy the curved rails directly from the manufacturer. All curves should be of sufficient radius for easy movement of cars and locomotives. Heavy locomotives with rigid wheel bases require curves of long radius.

Since the wheels of cars and locomotives bind on sharp curves, the clearance between the rails and the wheel flanges must be increased. The exact amount of the increase in gage width depends on curve radius, track gage and the wheel bases of the rolling stock, and no rule will apply to all cases. The width of the tread of the wheels limits the practical extra gage width.

Simple circular curves fill all speed requirements of underground tracks, and easement or transition curves are unnecessary. Degree of curvature is the angle at the center subtended by a chord of 30.48m, but curves of less than 60.96m radius are best designated by length of radius R.

Curves are laid out with a transit, or in various ways requiring no instrument work. A reasonably accurate curve layout can be made without a transit by means of offsets. Minimum curve radius is limited by wheel base, wheel diameter and flange clearance. Fig. 3-5, from an empirical formula, gives safe values for curves of same gage as on straight track. With liberal spread for gage on curves, 10%—25% may be added to wheel-base dimensions.

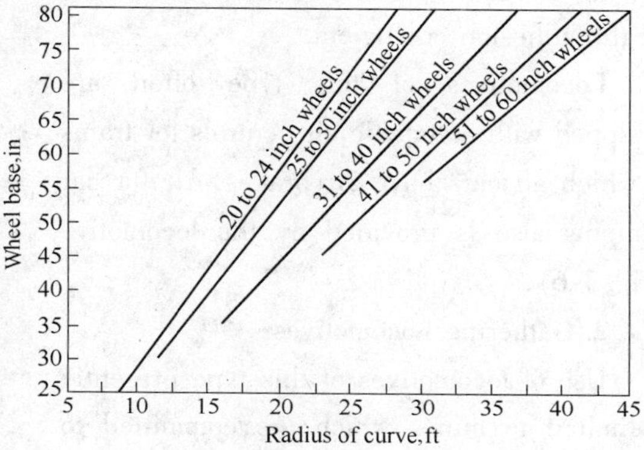

Fig. 3-5 Minimum-radius curve

Note: 1in = 25.4mm; 1ft = 304.8mm

3. Bending Rails

The common rail bender is made in four sizes for different weights of rail. Difference in length of inner and outer rails, D = gage × length of curve ÷ radius of curve.

3.1.2 Rail-Haulage System

3.1.2.1 Electric Haulage Locomotive Control

Modern mine haulage locomotives in the larger sizes normally are equipped with straight-parallel full electropneumatic controls for the motors. Dynamic braking is incorporated where long and/or heavy grades in favor of the loads are encountered. In the interest of safety, the braking-control circuit is operated from batteries permitting dynamic braking in the event the locomotive loses its power from the trolley. Other safety-oriented items included, in addition to air brakes, are air-operated sanders, trolley retrievers, uncouplers (in connection with automatic couplers), low-pressure-air warning devices and low-voltage headlights operating from the battery.

Silicon-controlled-rectifier (SCR) control for 250V dc motors of approximately 15kW has functioned satisfactorily on coal mine haulage equipment. Original and maintenance costs (plus a limited market) at the moment seem to be militating against its becoming a standard item. SCR equipment for large dc locomotive motors drawing approximately 3,000 amp per haulage unit have not found general use in the mining industry.

3.1.2.2 Special Locomotives

1. Locomotives for Metal Mines

Locomotives for this service usually are smaller than those used in coal mines because of limited dimensions of the openings. When built for "wet" service, they are equipped with closed cab and closely fitted covers over the top of the frame for protection of the operator and equipment. When estimating the weight of a locomotive for wet metal mines, the damp and slippery surface of rails must be considered in choosing a value for the adhesion coefficient.

Locomotives of this type often are equipped with automatic air controls for trains in which all cars have air brakes. Air for car dumping also is provided by the locomotive (Fig. 3-6).

Fig. 3-6 15t metal-mine haulage locomotive (protective covers and closed cab), straight-parallel semielectropneumatic contactor control and automatic air-brake control for train. Locomotive also provides air for car-dumping

2. Gathering Locomotives

Use of locomotives of this type presently is limited to mines which are committed to track haulage in the working sections. They are equipped with both a trolley pole and a reel on which a trailing cable is wound and unwound automatically, and which is connected to the electric-motor circuits by collecting brushes. Trolley nips and ground hooks on the outbye end permit connection to the power source in a fresh-air opening when the unit travels beyond such opening. Federal and state legislation and regulations require the use of explosion-tested equipment in gassy mines and restrict the use of these locomotives in certain mine areas. Gathering locomotives usually are built in 4, 6 and 8t sizes. Motor equipment is rated at approximately 7,340 W per metric ton locomotive weight.

3. Battery Locomotives

Battery-operated haulage locomotives are practically nonexistent in modern mines in the U.S. A number of larger units with trailers carrying the battery were operated by several mining companies in gassy mines until quite recently.

Locomotives with combination of trolley and battery power recently have been supplied for tunnel operations.

The U. S. Bureau of Mines issues approvals to cover battery-operated units which meet the construction regulations of its Schedule 2.

Battery locomotives are classed according to the chassis weight (without battery) and have been built in sizes from 1,361kg with one 4,474W motors and a 40V 14.5kW · hr battery, to 13,608kg with two 44,742W motors and a 250V battery with a capacity of approximately

130kW · hr. Motors are not designed to slip the wheels because of the extra battery weight. The locomotive weight, motor and battery are selected to meet requirements of a specific duty cycle.

3.1.2.3 Computer Applications

Computer programs for solving various mining problems currently in use include "Simulator Programs for Rail-Haulage Systems". A thorough knowledge of the details of both the program involved and the problem to be analyzed is essential for intelligent application of any computer program. Results obtained by means of simulator programs should be evaluated in light of the fact that simulation is only an approximation of reality. It is recommended that current literature be searched for latest developments in computer applications to mining problems at the time of need.

3.1.3 Trucks and Tractor-Trailers

3.1.3.1 Ore Haulage

Rubber-tired trucks used for underground ore haulage presently take various forms. Standard diesel highway trucks of relatively small capacity are quite common in certain parts of the U. S. They are used with room-and-pillar or other systems where opening height is sufficient and no headroom problems are encountered. These units generally are adapted for underground use by equipping them with exhaust scrubbers and, occasionally, adding or removing equipment to make them more suitable for working below ground.

Diesel units designed specifically for use in mines have been adopted in mining operations in recent years. These trucks are designed with a low profile for clearance in mines which cannot accommodate standard highway units. More important, however, is articulated design. The diesel power unit and the operator occupy one section of the unit and are separated from the ore-carrying compartment by a hinged joint. The joint is the basis for vehicle steering and also can absorb loading shocks so that the operator does not feel their full effect. The articulated steering provides more maneuverability and ability to make turns in tight places. These trucks can be used with loading units of many types, but most often are employed with diesel front-end loaders. Two types of interest are the Kiruna truck and the Wagner Teletram.

Combination vehicles, generally referred to as load-haul-dump machines, are finding extensive use in underground ore haulage.

While diesel-powered vehicles are making headway in metal and nonmetallic mining, recent advances in battery packs have been reflected in increased use of battery-powered units in coal mining. The units generally are referred to as "tractor-trailers." Basically, they are a compromise between a truck and a shuttle car and are popular because they have eliminated the restrictions imposed by a trailing cable. In addition, because the units can feasibly haul up to 610m, they can serve for both face and secondary haulage, eliminating entirely the butt-or section-entry belt in many mines. One source indicates that two tractor-trailers can haul up to 213m; three are required from 213 to 426m; and four for 426 to 610m.

Tractor-trailer units are of two general types. Kersey units (Fig. 3-7) have true trailers which

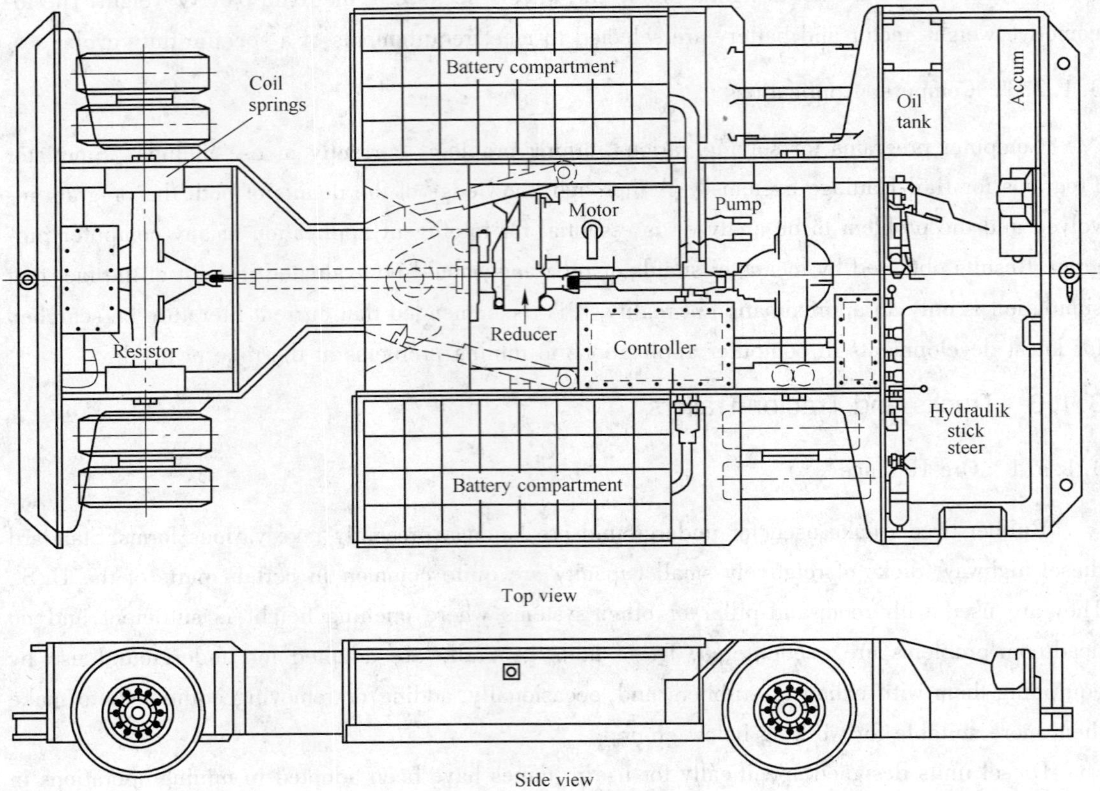

Fig. 3-7 Battery-power mine tractor

can be unhooked from the tractor. The trailers are end-dump units requiring a dumping station, which must be shot out of the bottom. In unloading, the operator backs the trailer down a ramp which automatically causes the trailer bed to tip and discharge its load. For greater flexibility, dual articulated tractors have been developed by S & S Machinery Sales. The second type of battery-powered vehicle is not a true tractor-trailer but instead is an articulated vehicle with a swivel joint between the tractor and trailer portions of the unit. These vehicles are manufactured by Galis, Jeffrey (Fig. 3-8) and Elkhorn, and all feature positive unloading, using a flight conveyor (Galis) or hydraulic rams (Jeffrey and Elkhorn).

While battery-powered tractor-trailers have their advantages, they all require a battery-charging station permitting rapid battery charge. In addition, several batteries must be kept on hand for each tractor so that fresh batteries always are available when needed.

3.1.3.2 Supply Haulage

The transporting of supplies to mining faces often is performed by trucklike vehicles. Where specific supply vehicles are used, battery-powered cars predominate in coal and diesel machines in noncoal operations. Diesel trucks and other similar vehicles are standard supply vehicles in many underground mines. This is especially true in mines operating in horizontal seams without a track

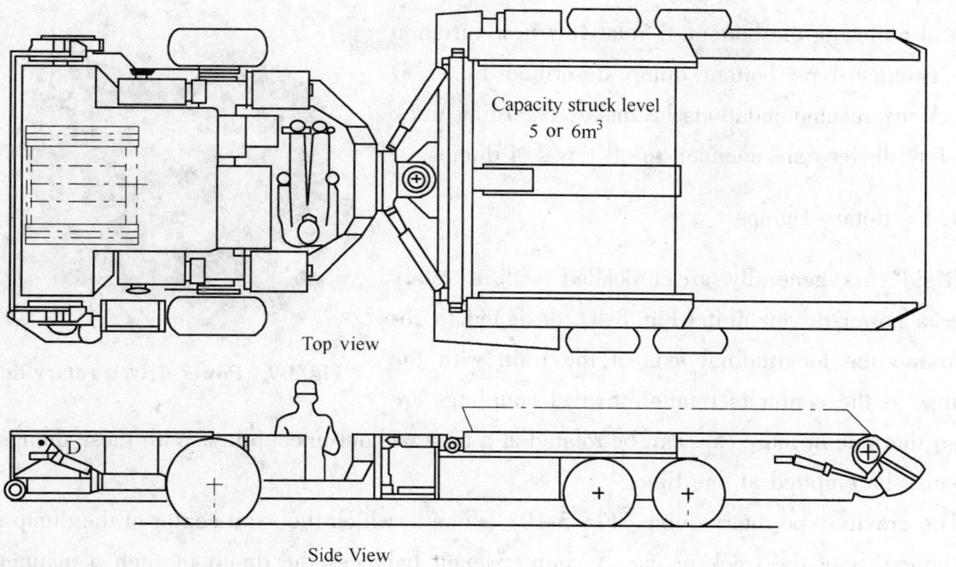

Fig. 3-8 Articulated tractor-trailer unit

haulage system. However, diesel engines meeting Schedule 24 of the USBM (the U. S. Bureau of Mines) are the usual power units.

Battery-powered trucks for supply and personnel movement are finding more acceptance in coal mines, particularly where each section is serviced by secondary belt haulage. Being rubber-tired, these vehicles are versatile due to their freedom of movement and the wide variety of utility jobs they can handle. Supply cars pulled by tractors are particularly versatile since trailers for many different purposes can all be hauled by a single tractor. In addition, several methods have been devised for transporting supply trailers on the main track-haulage system. These employ either dual sets of wheels for rail and roadway use or track dollies to carry supply trailers from the portal to the inby end of the track-haulage system.

3.1.4 Underground Dumps

Dumping of ore in a mine may be done by a shuttle car discharging onto a belt or into mine cars where material is transferred from face to secondary haulage or from secondary to main-line haulage. The material is removed with an elevating conveyor built into the shuttle car. A surge car or, sometimes, a feeder-breaker often is used between a shuttle car and a belt.

Dumping also may occur at a transfer point where a belt conveyor discharges its load onto another conveyor or into mine cars, or even into skips at hoisting areas. Dumping or discharging from belts underground usually is over the head pulley of the belt, though sometimes, where there are large storage areas, such as a shaft bottom, a tripper may be used to spread the ore uniformly. When dumping into cars, means are provided to maintain continuous haulage as nearly as possible.

Mine cars, at transfer points, may be dumped into bins, onto belts, or even into other cars. Bottom-dump cars require a bottom latch tripper and a means of reclosing the bottom. A relatively new variation is the Swedish-type bottom dump described by C. D Mann. Many recommendations he makes regarding track and safety devices are common to all types of dumps.

Fig. 3-9 Power-driven rotary dump

3.1.4.1 Rotary Dumps

Rigid cars generally are unloaded with a rotary dump—a power-driven unit (Fig. 3-9) designed to rotate around the longitudinal axis of the train with the couplings as the center of rotation. Swivel couplings are used so that one or more cars can be rotated at a time without uncoupling. With these dumps, whole trains may be emptied at one time.

The gravity-type rotary dump (Fig. 3-10) is one in which the axial center of the dump is offset from the center of the track or car. A counterweight balances the dump in such a manner that a loaded car will be top heavy and overturn. The counterweight then returns the car to its original position. With the gravity-type rotary dump, cars must be uncoupled and dumped one at a time. Automatic couplings often are used to ease and speed the process of coupling and uncoupling.

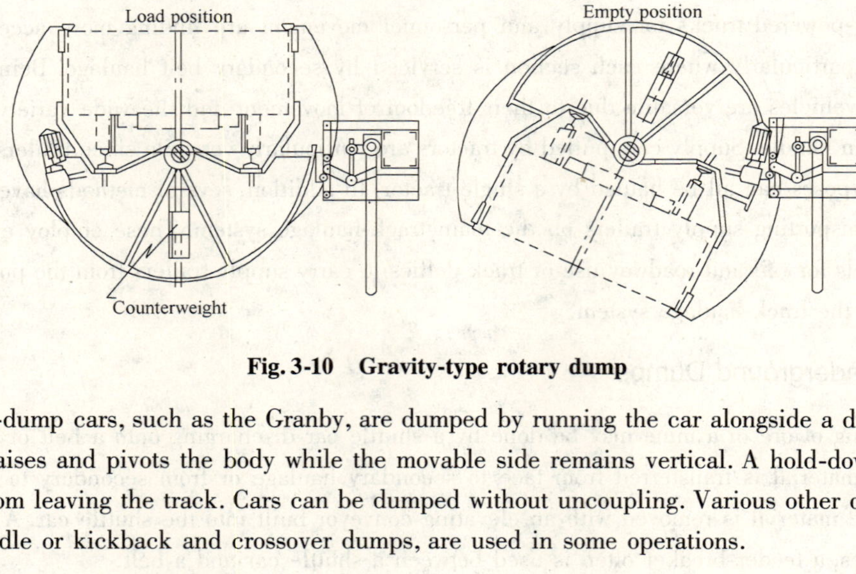

Fig. 3-10 Gravity-type rotary dump

Side-dump cars, such as the Granby, are dumped by running the car alongside a dump block. A roller raises and pivots the body while the movable side remains vertical. A hold-down prevents the car from leaving the track. Cars can be dumped without uncoupling. Various other dumps, such as the cradle or kickback and crossover dumps, are used in some operations.

3.1.4.2 Swedish-Type Car Dumps

1. Mine Cars

Design of mine cars depends upon the rate of production required, and involves a complete analysis of loading, traveling and dumping rates, haulage distances, etc. When the capacity of the mine car has been determined, special attention must be given to design of specific features of tops and bottoms.

Mine cars designed for continuous dumping by the Swedish System are bottom-dumping with two variations: where the bottom is hinged at (1) the end of the car (Fig. 3-11), and (2) the side of the car. In each case, the hinge pin or pins are the only positive connection between the top and bottom of the car. An extra wheel under the bottom, which bears the weight of the ore plus the car bottom, rides across the dump on a special monorail. Bearing wear in this extra wheel can cause deviation from allowable tolerances, which in turn can exert severe stresses on car bottoms as the car enters the dump and the load is transferred from the trucks on the haulage rails to the single wheel on the monorail. Damage to the monorail also can result.

Car bottoms must be designed to withstand stresses due mainly to the weight of the ore. Stresses are most severe midway between the hinge pin and the single wheel. In some larger cars, this span is a considerable distance. Car tops must be of sturdy construction, with design again dependent upon ore characteristics, as well as the special consideration that in this type of car the top must take all coupling and draw-bar strains. At precise locations on the sides of the cars, special beams are attached which ride on rollers during the dumping operation.

Mine-car tolerances must be small, as little as 3mm in the trucks, side rails or other massive sections which also are deformation-resisting assemblies. The tremendous force generated by the train as it accelerates and decelerates while unloading could damage cars or the dump if a car should catch on rollers, rerailers, or liner plates.

2. Trackwork

Leading to the dump, the track grade normally should favor the load, leading into the dump at no more than 0.5%. Track should be well ballasted and maintained. At the dump, track must be at the proper elevation with respect to dump rollers and monorail. Grade across the dump should be level.

3. Operation

As the train, locomotive first, approaches the dump, it always should be under control and prepared to stop. It may be necessary to obtain clearance from below before dumping or to visually inspect the location of ore in the dump. If ore is at or near the scroll, considerable damage can be done to car bottoms or trucks if dumping is attempted. The motorman may operate the locomotive and retarder from a control console at the side of the dump or ride the locomotive through the dump, controlling his speed with the retarder. At the exit end the locomotive rides onto the main track, picks up trolley power, and assists in pulling the empty cars out of the dump. Dumping can be accomplished at rates in excess of 127t/min. Special care must be taken with wet, sticky material, which can build up on the monorail or support steel and interfere with dumping. If a car does not empty completely, the remaining material can wedge between the car top and bottom as the train moves through the dump, putting tremendous stresses on the hinge and car bottom as it closes.

4. Auxiliary Devices

An I-beam monorail supporting a traveling hoist can be installed on the center line of the dump to hoist disabled cars or components from the dump, or for materials handling during dump repairs. Dust collection may be necessary. At MacLeod mine, a raise was driven from below the dump to connect it with a dust collector consisting of a fan and filter.

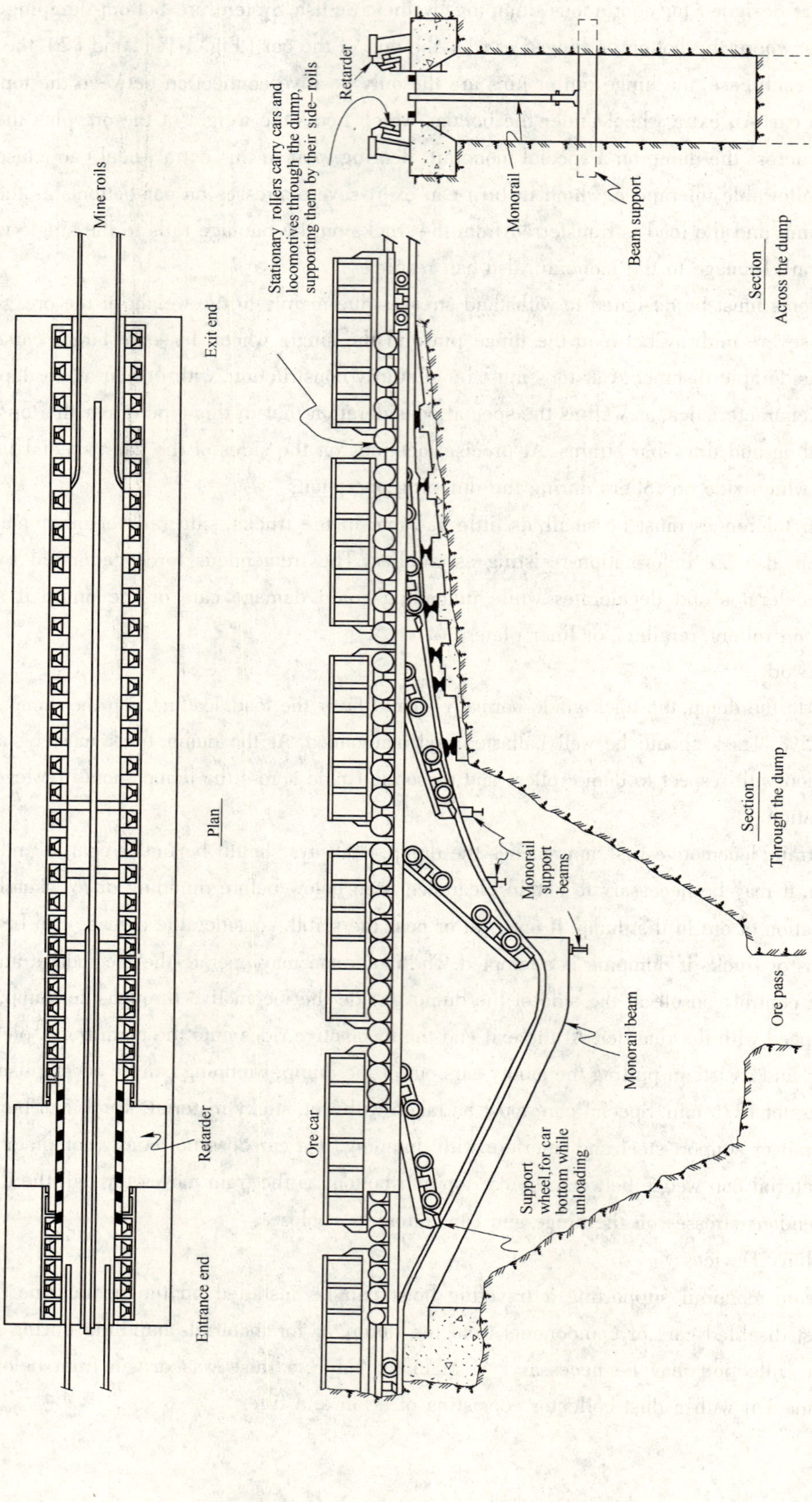

Fig. 3-11 Swedish-type car dump

3.1.5 Transportation of Men Underground

3.1.5.1 General Considerations

An important phase of underground haulage is safe efficient deployment of miners to their working places. As equipment becomes more expensive and potentially more productive, and as labor becomes more costly, it becomes increasingly important to minimize unprofitable and fatiguing travel time, thereby providing more actual productive time in the working place.

A recent survey of a large number of mines in the United States disclosed that travel time, exclusive of travel in shafts and slopes, averages approximately 48 min, or 10% of the normal workday. The requirement for shortest practical man-riding distances should be carefully considered in the design of the original underground layout and in subsequent mine-development planning. The main shaft or drift should be as close as possible to the main body of ore to be mined. In the case of coal or other minerals occurring in stratified beds, it is desirable to enter the bed near the geographical center of the boundary. Even so, some mines are so widespread that travel time is excessive. In such cases, where terrain features and ownership considerations permit, ancillary portal facilities should be considered. Such an installation is justified whenever cost analysis, including labor savings, shows an adequate return on investment.

The following information pertains to man-riding from shaft or slope bottom or from drift portal only.

3.1.5.2 Factors in Design of Personnel-Carrying Facilities

The basic design of man-riding vehicles will be governed largely by the physical characteristics of the mine. Widths and heights will depend on the dimensions of the entries to be traveled and curves to be negotiated. Speeds will be governed by condition of roadbed or roadways and grades encountered. Capacities will depend on size of crews to be transported to a given location. Braking capacity must be ample to provide a safe stop on the steepest grades expected, and towed vehicles should be provided with a device to prevent uncoupling.

Provision must be made for three basic types of man-riding: transportation of crews between portal and working face at the beginning and end of the shift; transportation of supervisors, inspectors and maintenance personnel during shift; and evacuation of persons or crew whenever required. Safety must be recognized as a primary consideration in any form of man-riding. In transportation of crews, the secondary consideration is rapidity. A minimum of travel time produces maximum utilization of expensive production equipment and minimum out-of-pocket expense for nonproductive labor. Supervisors, inspectors and mechanics need reasonably quick transportation, but travel during the shift must be accomplished without interference with the flow of product or supplies. When transporting injured persons, the emphasis is directed toward keeping them as comfortable as possible and moving them in a way that will not aggravate injuries. All man conveyances should be designed to accommodate a stretcher.

3.1.5.3 Mines with Rail-Type Main and Secondary Haulage

1. Vehicles

Three types of vehicles are available for transportation of men in rail haulage systems. In many mines, pit cars other than the drop-bottom type are used. They have the disadvantages of uncomfortable seating and no protection from spalling roof or water drips, and generally are not equipped with safety chains or other desirable safety features. Also, unless a certain contingent is reserved, they are not always available when needed. An alternative is specially designed man-riding cars for use with the system' s locomotives. These cars are available in a wide range of sizes to accommodate from 8 up to 40 men. Fig. 3-12 is a configuration drawing of a 36-man car. They usually are covered and equipped with automatic couplings, plus safety chains between cars and between car and locomotive. An emergency signaling cord sometimes is provided. Separate compartments are provided for fire extinguishers and rescue breathing apparatus, where required, or for small tools or supply items. Cars also should be equipped with restraining chains across access openings and, in the case of cars used with trolley locomotives, an insulated roof.

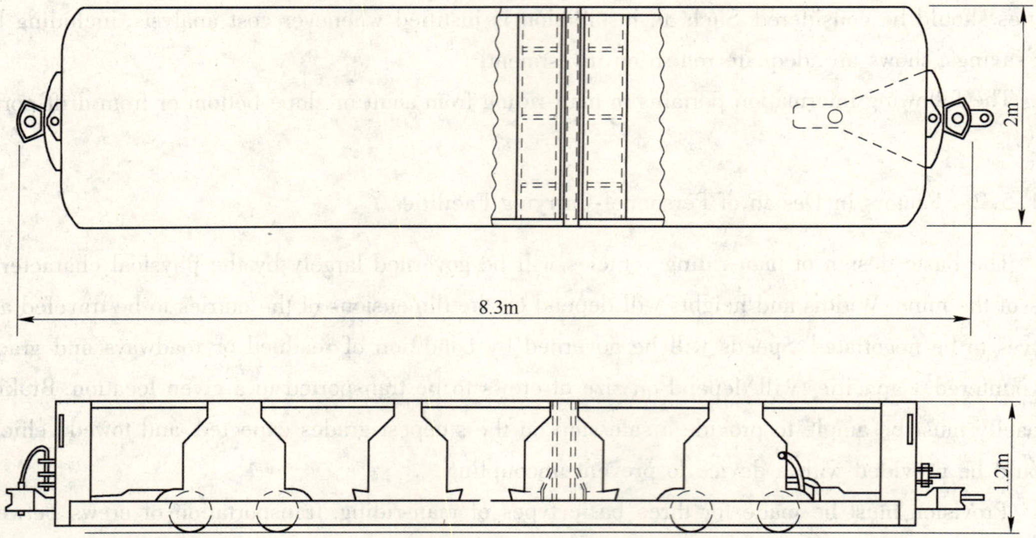

Fig. 3-12 Typical 36-passenger personnel car—swivel truck, 305mm wheels, designed to negotiate 15m-radius curves, safety-chain hitches, brakes applicable to all four wheels, 5mm-steel roof with 3mm of insulating material

Mines using man trains of pit cars or special man-trip cars find it desirable to include a small complement of self-propelled rail vehicles for on-shift travel. Some mines use self-propelled units for all man-riding, usually where working places are so scattered that the man-train route would be unnecessarily circuitous. These units have the advantage of being slightly faster than man trains and more versatile. They can be used for on-shift travel and delivery of spares or supplies. If parked near the working area, they are quickly available for evacuation or ambulance duty. A wide variety of these units is manufactured. They may be either open or covered and generally feature, as standard equipment, two independent braking systems, self-centering controls and insulated roofs. The open type is shown in Fig. 3-13; closed, Fig. 3-14.

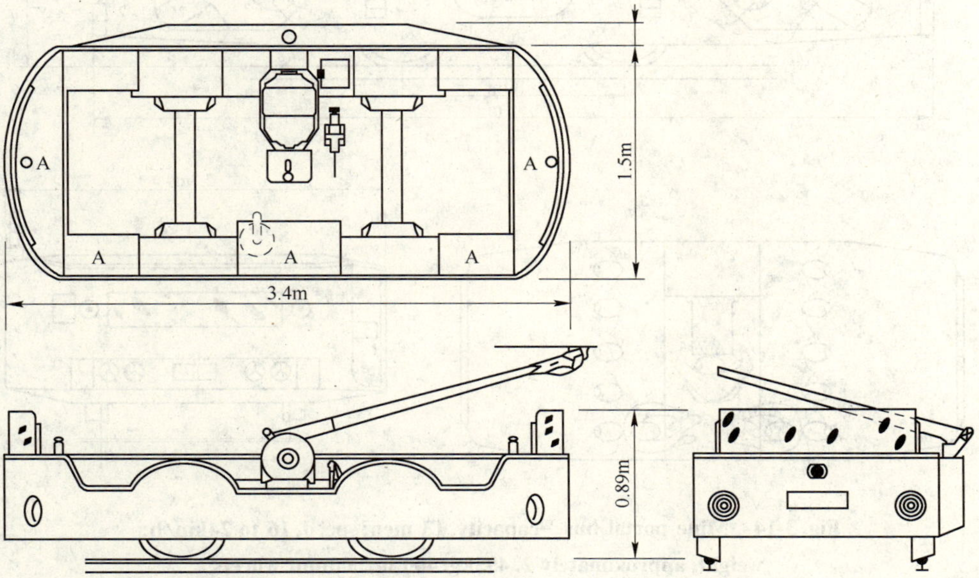

Fig. 3-13 Mine Jitney—capacity, 7 men; speed, 13 to 16km/h; empty weight, approximately 1,360kg; width, 1650mm for 910 to 1220mm track gage or 1900mm for 1440mm gage; 5.6kW wound-rotor motor; 400 or 500mm wheels; optional seats marked "A"

2. Cost of Equipment

Due to the extremely wide variety of pit cars in use, it is not considered useful to estimate prices. Specially designed man-cars range in price from $ 2,000 for small units up to $ 6,000 for cars seating 40 men. Self-propelled vehicles are available from $ 4,500 in the four-man size up to $ 17,500 for a 24man car.

3. Speeds Attained

A tabulation of information received from a large number of metal, coal and other nonmetallic mines reveals a wide range of man-riding speeds. For rail man trains, the range was 4.8 to 21.9km/h, with the average 12.6. Though speeds are limited by law in some localities, most operators reported rates considerably lower than the legal limit, and cited grades and roadbed condition as the limiting factors. A tabulation of mines using self-propelled rail equipment showed an average of 13.4km/h with a range of 4.8 to 24km/h.

4. Other Considerations

Sound planning for positive traffic control is vitally important in the design and operation of any rail system. Planning must be directed toward maximum safety, and man trips should operate

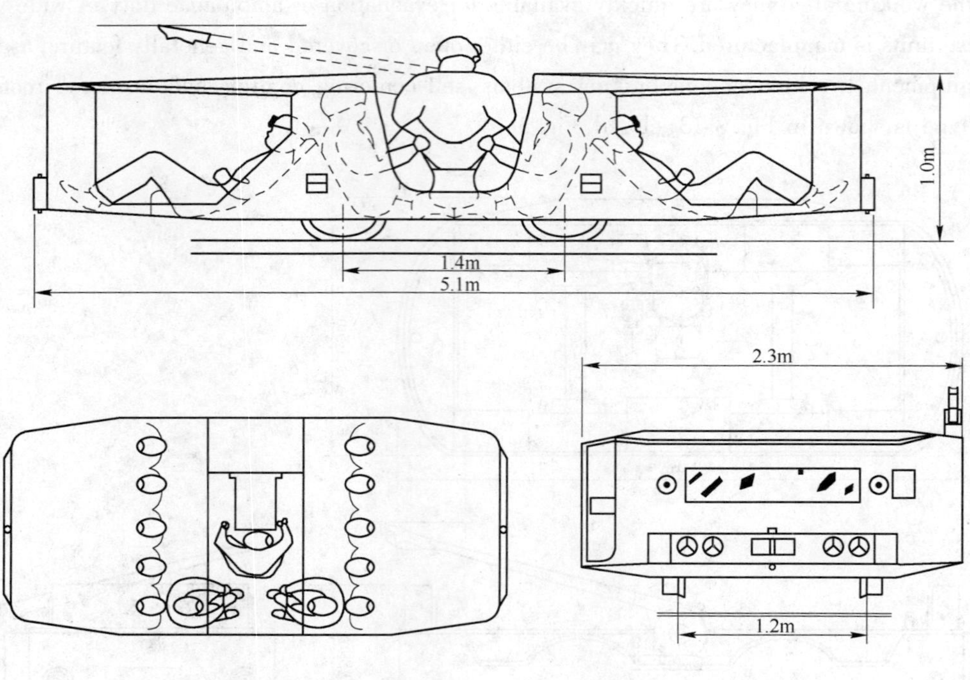

Fig. 3-14 Mine portal bus—capacity, 13 men; speed, 16 to 24km/h; weight, approximately 2, 495kg; 400 or 500mm wheels

under a signal block system and/or control of a dispatcher.

3.1.5.4 Mines Utilizing Rubber-Tired Haulage

In some mines where internal combustion equipment is permitted, off-track transportation is the predominant mode. In such cases, operators of mobile equipment usually find their units parked at the shaft bottom and drive them to the active workings. Many operators add small truck beds to farm-type tractors and use these to transport men and distribute explosives and other supplies. When used for transporting men, seats, canopy and restraining gates should be provided. Small diesel-powered jeep-type vehicles and/or diesel pickup trucks are in common use. Articulated vehicles can be purchased with several sizes of man-riding bodies (Fig. 3-15).

1. Cost of Vehicles

Many man-riding vehicles are shop-made at the mines by addition of a bed or platform to existing equipment. Farm-type tractors equipped with exhaust scrubbers range from approximately \$3, 500 up. The price of jeep-type vehicles, similarly equipped, ranges from \$4, 000 to \$7, 000. The articulated vehicle shown in Fig. 3-15 is available for approximately \$14, 000.

2. Speeds Attained

Speeds of rubber-tired equipment for man-riding are almost entirely governed by grades and roadway condition, since internal combustion equipment usually is capable of speeds well in excess of those desired underground. Mines responding to inquiries reported an average of 10. 6km/h with

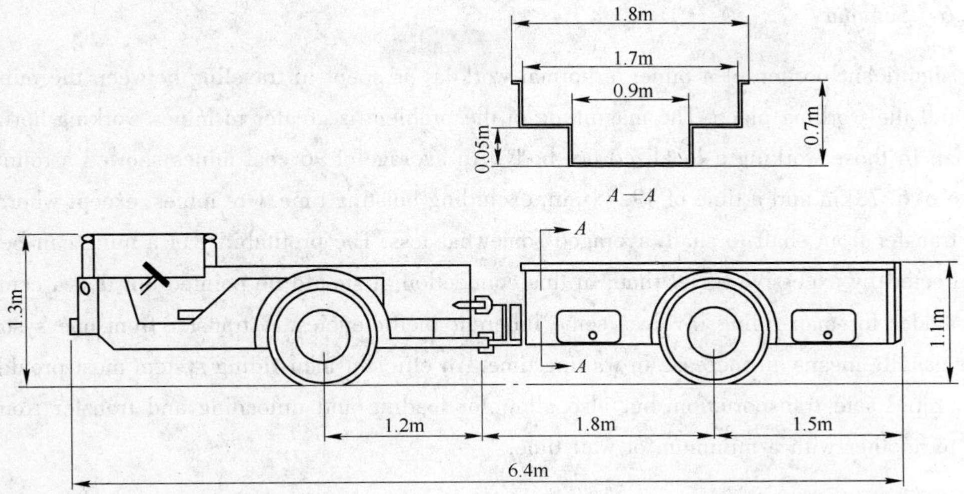

Fig. 3-15 Personnel carrier—capacity, 14 men; 4-cylinder air-cooled diesel engine, 31.3 gross kW; travel speed, 6.4 to 24km/h forward and 3 mph reverse; 8:25 × 15 tires; 9ft 6in. inside and 17ft 4in. outside turning radius; weight, 8,764 lb

the range 4.8 to 19.2km/h.

3. Other Considerations

Special attention should be paid to roadways for off-track transportation. With rising labor costs, a considerable saving can be obtained by even small reductions in traveling time. Most operators using this type transportation find that keeping roadways well drained and graded also pays dividends in reduced maintenance costs. Some operations find it profitable to pave or apply metal to roadways in the more troublesome areas if they are to be used for a long period. The amount that can be profitably spent for roadway improvements depends on the life of the area and the number of men being transported.

3.1.5.5 Cost Comparisons

Cost comparisons for different transportation systems may be made by inserting the time factor into a simple unit-cost formula. In this method, hours paid but not available for productive work are extended by the wage rate and added to the cost of operation the man-riding equipment to arrive at the unit cost of man-riding:

$$\text{Cost per man-km} = \frac{(\text{equip. hr})(\text{cost per hr}) + (\text{avg. no. men})(\text{elaped time})(\text{wage rate})}{(\text{avg. no. men})(\text{km traveled})} \tag{3-1}$$

The cost per operating hour should include depreciation, taxes and all other items of expense relating to man-riding equipment.

3.1.5.6 Summary

A significant portion of a miner's normal workday is spent in traveling between the mine entrance and the working place. The magnitude of the problem is greater in mines working horizontal beds than in those working a localized ore body. An average of 96 coal mines showed a round-trip distance of 6.78km and a time of 49.74min, excluding hoisting time. Ore mines, except where men had to transfer from shaft to shaft, averaged somewhat less. The profitability of a mine can be seriously affected by excessive travel time. In this connection, it should be pointed out that a combination of modes for man-riding involves some inherent inefficiencies. A transfer from one system to another usually means an increase in waiting time. An efficient man-riding system must provide not only for rapid safe transportation, but also allow for loading and unloading and transfer from one system to another with a minimum of wait time.

3.2 Surface Haulage

3.2.1 Railroad Haulage

Conventional railroad haulage can be employed at both open-pit or strip and underground mines—in the latter case, largely for transfer of ore from distant openings to processing plants, or processed material to loading points or docks, using standard hopper cars for bottom unloading or gondolas for rotary dumping. The major application, however, is in surface mining for moving ore and/or waste out of the pit to a crushing plant or dump. Consequently, the following will deal largely with this phase of railroad operation, though the conclusions apply equally, for the most part, to other rail-type surface haulage.

In surface mining, trains consisting of a locomotive and gondola or side-dump cars suitable for loading by power shovels are used. Gondolas are restricted to hauling ore as a rotary dumper is required. Side-dump cars are more versatile, carrying both ore and waste, but because of their additional dumping mechanism they are likely to be more expensive to maintain. Also they weigh up to twice a flat-bottom car of the same capacity and, therefore, fewer cars can be hauled per train. Standard-gage track is most commonly used. Track systems incorporate low-quality temporary track where moves are frequent and permanent high-quality track on main lines. Rail haulage can be used alone, or in combination with other surface haulage methods, thus permitting a selection best adapted to the peculiar requirements of each mining project.

3.2.1.1 Applicability of Railroad Haulage in Open-Pit Mines

For handling rough rock, the shovel-train system is unexcelled. The ideal application is in a large, but not excessively deep, open pit mine from which the coarsely blasted ore and waste must be transported over 4.8km. If, in addition, a high production rate is to be maintained, railroad haulage probably will afford a lower cost per metric ton transported. Grades must be relatively gentle—usually 3% uphill and 4% downhill. The greater the differences in elevation involved, the greater are the track grade space requirements; for example, a 3% ramp out of a pit only 91m deep is almost 3.2km long.

Main-line track locations should be selected carefully so that the number of moves during the life of the mining operation can be kept to a minimum. This applies both to the ore haulageway and that for waste. Such moves are not only expensive, but are difficult to execute without adversely affecting production rates. The large mine affords greater opportunities for judicious location of semi-permanent trackage. Fig. 3-16 and Fig. 3-17 illustrate typical track layouts.

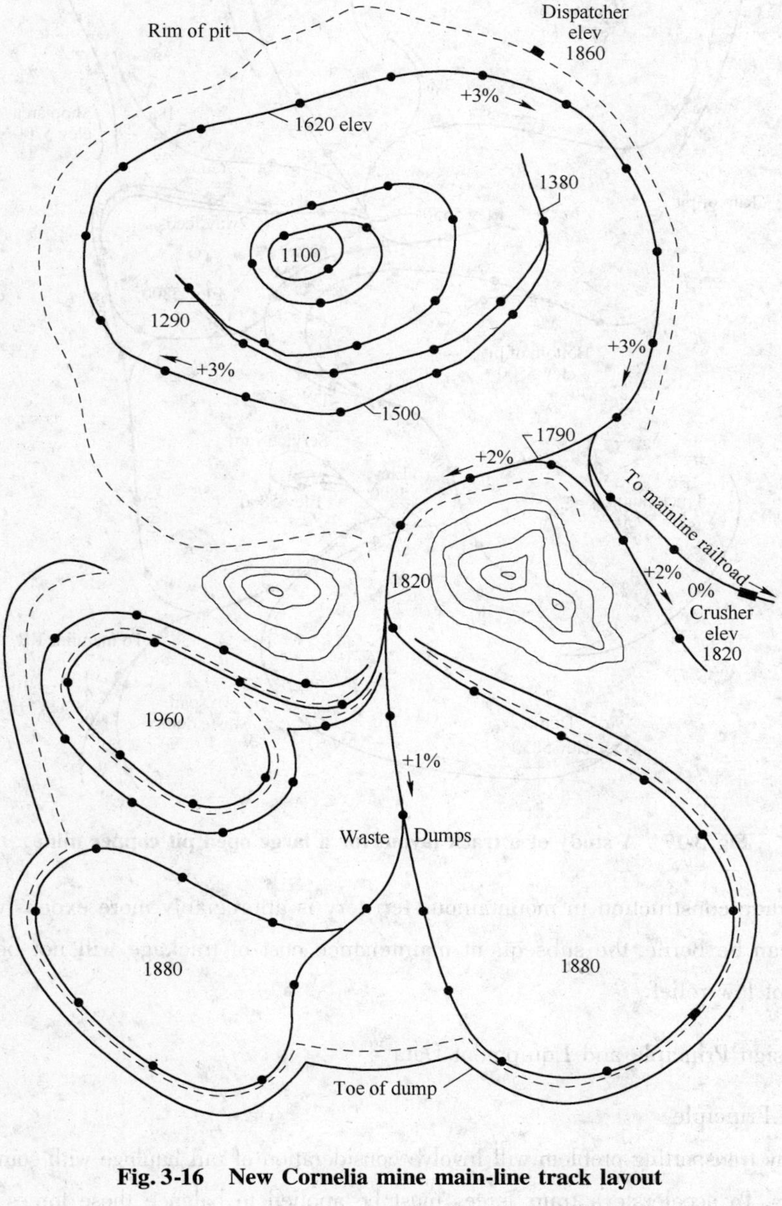

Fig. 3-16 New Cornelia mine main-line track layout

The effect of terrain in considering railroad haulage is confined chiefly to the initial cost of the

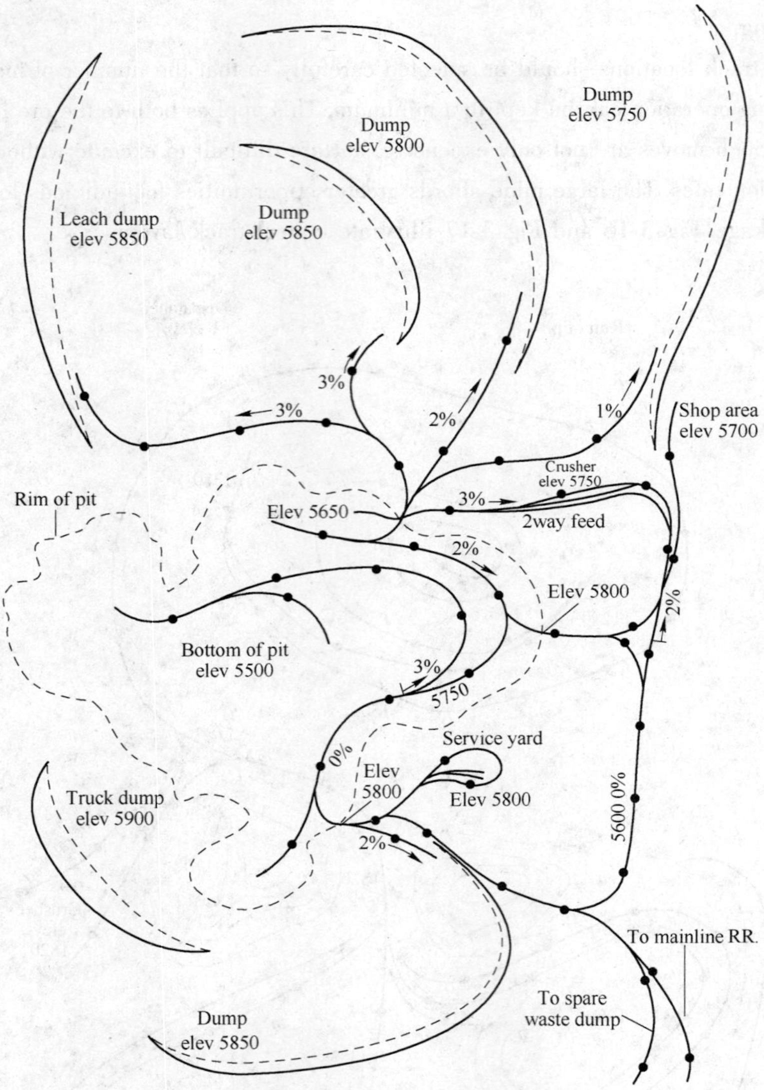

Fig. 3-17 A study of a track layout for a large open pit copper mine

installation. Grade construction in mountainous territory is appreciably more expensive. However, if that expense can be borne, the subsequent maintenance cost of trackage will not be much higher than in areas of low relief.

3.2.1.2 Design Principle and Equipment Data

1. Design Principle

Almost any transporting problem will involve consideration of rail haulage with some type of locomotive and cars. To accelerate a train, forces must be applied to balance those forces opposing train movement and to provide additional accelerating energy. These opposing forces are shown in Fig. 3-18.

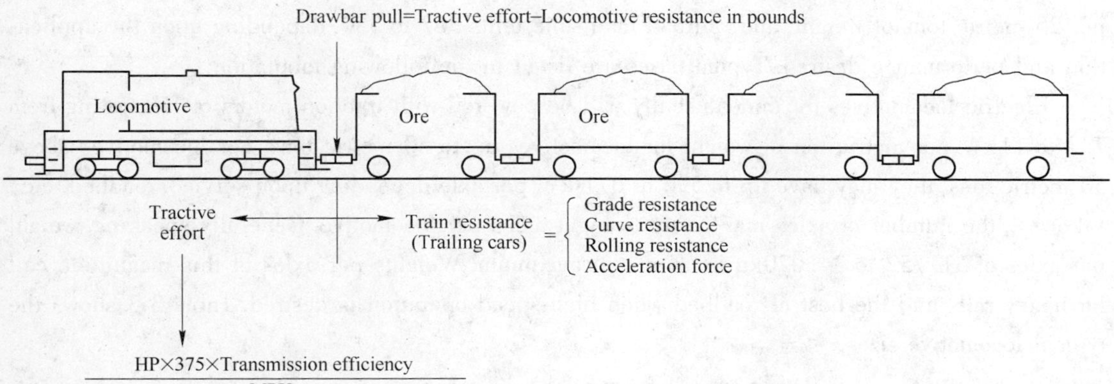

Fig. 3-18 Accelerating and resisting forces in railroad design

Note: 1lb = 0.4536kg; 1hp = 745.7W; 1mph = 1.6km/h

2. Equipment and Track

A locomotive having one or more diesel engines driving electric generators feeding traction motors is called a diesel-electric. The diesel-electric locomotive has some of the advantages of the electric locomotive in that transmission is highly efficient, maintenance costs are relatively low, availability is high, and production of standard-size units is relatively high. Furthermore, diesel engines generally have low operating costs. The diesel-electric transmission provides high tractive efforts for starting and accelerating heavy trains. A train consisting of 125 to 140 cars, 41 to 132 metric tons each, is quite common on many U. S. railroads. The rated horsepower of the diesel engine is the maximum available for traction. In railroad switching service, 447 to 1,342kW generally is ample, but in road service, where a high starting tractive effort is desirable and high-speed operation is essential, several locomotives are operated in tandem giving up to 4, 474 to 11, 186kWper train.

In most mining applications, the short-time rating of the electrical equipment (less than 1 hr at rated engine output) is what determines the capacity and sometimes the weight of the locomotives. The profile of the track, maximum grade encountered, number of cars, and the weight of the loaded cars will determine the weight of locomotive and, possibly, the ratings of the electrical equipment. Over a rolling profile (one with up and down grades), the maximum grade, if short, may be negotiated by using the momentum of the train, but such applications are the exceptions. So, it is considered that the length of grade is sufficient to cause the locomotive to slow down to a constant speed, and thus it must have ample tractive effort to pull the train over the grade without slipping its wheels. Slipping occurs when the tractive effort at the wheels exceeds the available adhesion between the wheels and rail. With clean dry rails, adhesion up to 40% can be obtained, but for ordinary conditions 30% is the maximum, and for all weather conditions, 20%.

Mining and industrial locomotives, either electric or diesel-electric, may be either two-, four- or six-axle design, usually with all axles motorized or driven.

Industrial and railroad diesel-electric (switching) locomotives usually will have about 110kW per 25 metric tons of weight, and railroad main-line units 367 to 550, depending upon the application and performance desired. Typical sizes are listed in the following tabulation:

Electric locomotives for mines usually will be powered with traction motors of 1 hr rating from 7.3 to 11kW per metric ton of weight for locomotives up to 50 metric tons. For locomotives above 50 metric tons, they may have up to 596 or 671 kW per axle depending upon service, roadbed, etc. Likewise, the number of axles may go up to eight axles per locomotive. Generally speaking, weight per axles of 31,752 to 34,020kg is about the maximum. Weights per axles of this magnitude call for heavy rails and the best of roadbed when high-speed operation is desired. Table 3-3 shows the typical locomotive sizes.

Table 3-3 Typical locomotive sizes

	Typical locomotive sizes							
Locomotive weight, t	23	41	73	91	113	113	163	172
Gross hp, kW	112	224	410	567	895	1,491	2,237	2,237
No. of axles	2	4	4	4	4	4	4	4

Industrial locomotives may be for gages from 457 to 1,676mm; railroad, 610 up to 1,676mm. Width of industrial and railroad locomotives may be track gage plus 635mm.

Track grades usually range up to 2% or 3% on railroads and may go as high as 8% or 9% for mine or industrial applications.

Track curves may be 889mm radius on industrial applications, and on commercial railroads, 152 to 213m may be found. Most railroads and some industrial users may raise the outside rail as much as 102mm to counteract the centrifugal force generated by the train in going around the curve at some safe speed. The centrifugal force is $F = WV/(gR)$, where W is the weight of car or locomotive; V, speed in m/s; R, curve radius in m; and g, acceleration due to gravity. The height of outside rail above inside is called "superelevation". Another way to reduce centrifugal force is to reduce train speeds on the curved track. A combination of superelevation and reduced speed often is used to counteract centrifugal forces on curves.

The power supply voltage for industrial electric locomotives may be 250, 500, 600 or 750 V dc for yard or track distances of 4.8 to 8km. For 6.4 to 16km, the voltage may be 750 to 1,500, but the initial cost of the overhead line and locomotives will be 10% to 20% higher for 1,500 V than for 750. For high traffic densities, voltages of 11,000 to 24,000 ac, using either simple or compound catenaries (overhead wires, hangers, contact wire, catenary and messenger wires) are used. When two, three, four, or more tracks are used in yards a cross catenary may be required to support the overhead wire system over these tracks. Supports for overhead power supply for electric locomotives can vary from a simple wooden pole, with support bar holding the insulator and contact wire, to the complicated catenary system mentioned previously with its necessary supporting structures.

3.2.1.3 Special Items Specific to Rail Haulage

1. Traffic-Control Systems

For two or more trains to operate safely on a single track, some means of regulating traffic are needed. Simple manually operated signals suffice where only two trains are involved, or in the case of one train operating under conditions of restricted speed and clearance, as when track is being repaired. Automatic block-signal systems are a very effective means of regulation. Block signals on a single track can control traffic in one direction only. Two-way traffic can be accommodated if passing tracks are located at convenient intervals, and for most efficient operation two tracks are required.

Many mine trains are equipped with two-way radios for communication between enginemen, supervisors and maintenance men. These radios also afford a means of traffic control.

Where train traffic is dense, a centralized traffic-control system (CTC) is superior. CTC is unequalled in its ability to handle trains rapidly and safely over a given stretch of track. The operator, working from a control panel containing a diagram of the tracks in the system, selects the route he wishes each train to follow. Lights on the panel indicate the positions and movements of the trains, as well as the aspects of the signals and positions of the switches. Signals either along the wayside, in the locomotive cab, or sometimes both, tell the engineman how a train is to proceed. CTC is the most expensive of the conventional traffic regulating systems.

2. Remote Control and Automatic Operation of Trains

Within recent years there have been revolutionary developments in the field of automatic train operation. Three types of systems have been developed as of this writing: remote control, supervisory control, and programmed control. At least two have been applied to surface mine railroads.

With radio-control systems, operators control all locomotive operations by radio commands from pack sets to receivers on the locomotives.

Wired remote control is more easily applied to electric locomotives. The power for propulsion is from an overhead wire, and the control is achieved by a second collector and another overhead wire. Power for wired induction control is obtained either from an insulated wire laid on the ground surface near the rail or from the overhead trolley.

Supervisory control involves dispatcher control of train movements between various points. For protection against conflicting moves, train safety facilities are built into the system, thus freeing the dispatcher for major operations only.

The programmed-control system involves no planned manual supervision. In some now in operation, a sequence of events is performed during a complete cycle while in others part of a cycle must be completed before the remainder can proceed. Provision is made for overriding so that an operator or maintenance man can correct a defect or stop operation at any time.

Programmed train control at Carol mine in Labrador is accomplished by coded ac transmitted through the rails to iron-cored coils on the locomotive and tail car of each train. These coils act as receivers and in turn actuate relays and electropneumatic valves to produce locomotive responses. The operation of four trains is automatically programmed so that as laded Train A leaves the mine, Train B has begun to load, Train D is being dumped at the crusher, and Train C is returning empty.

Much experimental work is being conducted with a view to improving automation of commuter trains. Undoubtedly this will also benefit rail haulage in making available to mine managers a choice of several different modes of automatic train haulage control.

3. Maintenance Facilities for Rolling Stock

Adequate and well-equipped repair shops for locomotives and cars are a wise investment. Locomotive jacks are necessary to permit removal and replacement of trucks. Other maintenance work on locomotives logically divides into two types: (1) that on brake systems, traction motors, etc., most easily performed at track level; and (2) that on the engine, generator, air compressor and other components atop the frame, best performed from the level of the frame or deck. Platforms at that level are a great convenience.

Minor repairs to ore cars can be made in the open, but major repairs and overhauls are best accomplished in a shop building designed for the purpose. Automatic welding equipment for body work is very desirable. For extensive body repair or rebuilding work, equipment for cutting and forming large steel plates is essential.

Track Maintenance—Track maintenance is vital to an efficient rail haulage system. Good alignment and surface on main-line tracks is necessary. Rough track not only slows trains but contributes greatly to damage to rolling stock, crossties and wear on rails. This normally is the largest element of rail haulage cost, sometimes amounting to as much as 50%. A number of track maintenance machines developed in recent years are very useful in pit track work, including tie tampers, power jacks, ballast distributors and shapers, track liners, tie spacers, spike drivers and bolt tighteners. Tampers and lining machines have been provided recently with controls capable of bringing track automatically to a preset grade and line.

Rail Waste Dumps—These are more expensive to establish and maintain than truck waste dumps. A grade must be prepared upon which track can be laid to start the dump. The initial track should be at least a train length long. Preferably it should be several train lengths.

Dump height is largely dictated by terrain. Dumps in mountainous areas are likely to be high, while those in areas of little relief generally low.

High dumps require less frequent track throws, but are more susceptible to settlement. Continued work therefore is necessary to prevent development of excessive grades.

Rail dumps routinely develop a berm along their crests as the dumper material builds out from the track. When the berm becomes sufficiently high to seriously impede sliding of the muck out of a dump car, the dump is said to be "plugged". It then is necessary to blade off the berm and shift the track out of the new crest.

Waste trains are commonly pushed onto dumps. This is a precaution to save the locomotive from the likelihood of derailment and from the rare possibility that the dump crest may slough. More derailments occur on dumps than in any other part of a pit track system. Track quality necessarily is only fair because the track receives considerable abuse from frequent moves and derailments. Material covers the outside rail as the plug develops, frequently causing cars to jump the tracks as they trail through it.

3.2.1.4 Unit Trains

The unit train is a dedicated train-set of cars moving on a regular schedule, both loaded and unloaded, from one origin to its destination. The origin and destination may not be the same on each trip but the train is moved intact both loaded and empty. The unit train may be handled by locomotives from a pool or from another railroad but usually the locomotive also is dedicated and stays with the cars.

The first true unit train involved a 1,280km round trip from a coal mine to the Martins Creek power plant of Pennsylvania Light & Power Co. Loading the seventy-four 102t hopper cars, owned by the company, is accomplished in approximately 3 hr, meaning adequate facilities at the mine and destination to make the train a most economical tool for the power company. Many unit trains are now hauling coal, iron ore, phosphate, grain and other types of bulk commodities.

Top loading is the rule in practically all unit train operations. There unloading for the most part employs a rotary dump or bottom-dumping self-cleaning hopper-type cars. Side-dump cars are not employed in unit train service.

The high-sided tight-bottomed gondola-type car with a rotary coupler at one end suitable for rotary dumping also is relatively light, less costly and more maintenance-free. The bottom-dumping self-cleaning hopper-type cars with powered doors which can be opened and closed as the train moves slowly over an undertrack pit permit unloading at rates up to 10,160t/h. When made of aluminum, weight is around 24t. Therefore it can carry 106t of coal without exceeding the rail-limit gross weight of 119t. However, its first cost in 1966 was about 65% higher than the gondola for rotary dumping.

Methods of opening bottom dump cars are:

(1) Electric motors actuated by a probe automatically inserted into a receiver in the car side, permitting six standing cars to be unloaded at one time;

(2) A gear train powered from a rubber-tired wheel on the side of the car contacting a hydraulically elevated ramp rail, thus opening the doors when coming into the pit area and closing them when leaving;

(3) A horizontal-axis, rubber-tired wheel at the middle of the car underside contacting a ramp rail to open and close the doors.

Alternatives to locomotive movement of cars while loading coal or other materials through overhead chutes are the car haul, the tripper conveyor and the hydraulic ram-type car pusher.

3.2.1.5 Computer Use

The digital computer is being used to obtain approximate train performance, locomotive tonnage ratings and consumption of fuel or electrical energy. A locomotive speed-retractive effort curve is used in determining the locomotive or set of locomotives to be used to haul a train over a given profile. Using the Davis equations, the resistance is calculated for the train under study. Total train resistance, plus any power required for operating axle-driven auxiliaries, is subtracted from the locomotive speed-tractive effort curve of the locomotive in question to arrive at the net tractive effort available to produce acceleration on level tangent track. In computer language, dividing net tractive effort by 100 times total train weight in metric tons give the maximum acceleration on level tangent track in m/s^2 (minus for retardation), abbreviated as M. A. L. T. and called "MALT".

Three sets of punched input cards are used:

(1) The basic "programs", or lists of instructions (about 400 cards).

(2) Train Malt data calculated beforehand (five cards).

(3) Track data (grades, curves and km posts, maximum speed and speed restrictions, rate of deceleration during braking, and number, location and duration of stops).

One punched card is used for every point of change in track data and supplies the data ahead of it to the point represented by the next card. The beginning of the braking zone also must be specified on the card. The acceleration calculations, as done by the computer, take into consideration the change in Malt for each change of speed, grade and curve. A code is used to identify the operating conditions, such as starting, full throttle, entering slow zone, entering stopping zone, and the time at timing points.

3.2.2 Truck Haulage

In the ten years prior to 1972, open pit mining has seen the evolution of haulage trucks from twin-engine mechanical-drive 36-t units to single-engine electric-drive 254-t units. The ability of trucks to operate despite rapidly changing pit outlines and to handle multi-product hauls cannot be approached by rail systems. As a consequence, truck haulage has become the single most important cost element in open pit mining.

3.2.2.1 Haulage Unit Types and Equipment

1. Rear Dumps

These units have the body mounted on the truck frame, and are dumped by raising the box with a hydraulic-hoist system. As the most-common type of truck, they are capable of handling all types of material, whether blasted, ripped, or loose. Fig. 3-19 shows four-wheel vs. six-wheel "load-on-back" rear-dump trucks. Fig. 3-20 gives overhung vs. conventional two-axle tractor with "rocker"-type rear-dump trailers.

(1) The two-axle truck usually is driven only by the rear wheels, although four-wheel drive units are available. Dual tires are standard on the rear axle, and also may be used on the front axle.

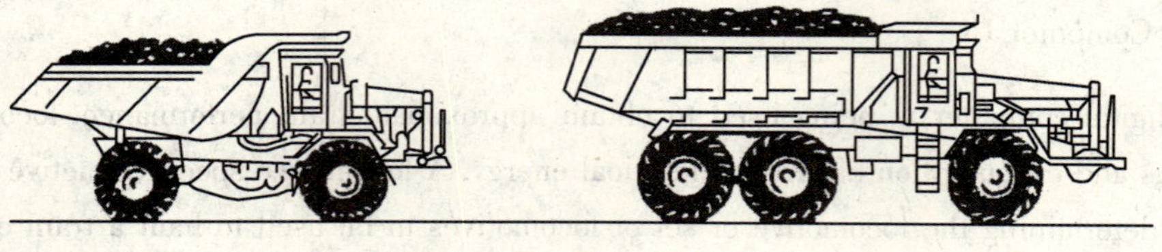

(a) (b)

Fig. 3-19 Four-wheel vs. six-wheel "load-on-back" rear-dump trucks

(a) (b)

Fig. 3-20 Overhung vs. conventional two-axle tractor with "rocker"-type rear-dump trailers

(2) On the three-axle truck, both rear axles drive and dual tires are standard on the rear axles.

2. Trailer Dumps

Since the rock-carrying portion of these units is detachable, they have the advantage that the entire unit need not be backed up by a complete standby unit, but perhaps only the prime mover.

(1) Since the single-axle prime mover can turn 90° in either direction without any forward motion, this unit is very maneuverable. In quarries, tunnels or where narrow roads are required, this can be a great operating advantage.

(2) The two-axle tractor prime mover does not have the maneuverability of the single-axle unit but, since the ride characteristics are much better, it is more suitable for long hauls. Fig. 3-21 shows side-dump tractor-trailer unit with option of either integral hydraulic or external "sky-hook" hoisting arrangement. Fig. 3-22 gives bottom-dump tractor-trailer unit with either drop or clamshell doors.

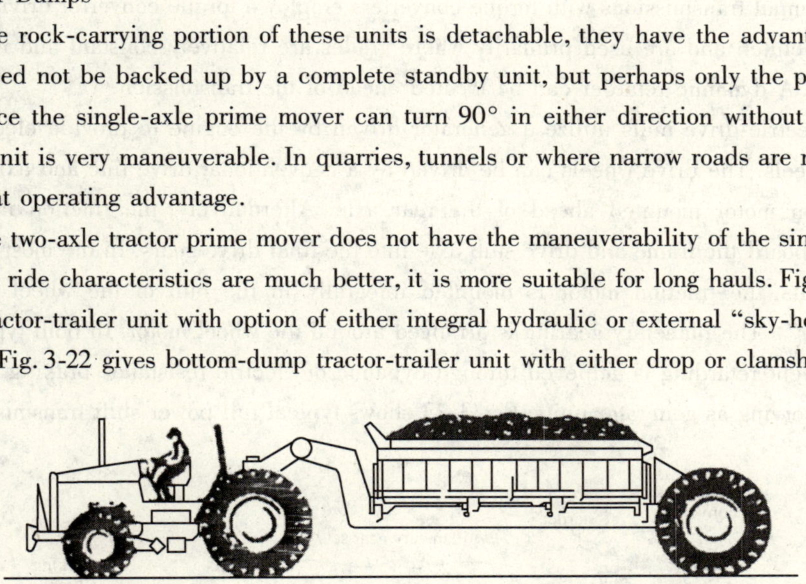

Fig. 3-21 Side-dump tractor-trailer unit with option of either integral hydraulic or external "sky-hook" hoisting arrangement

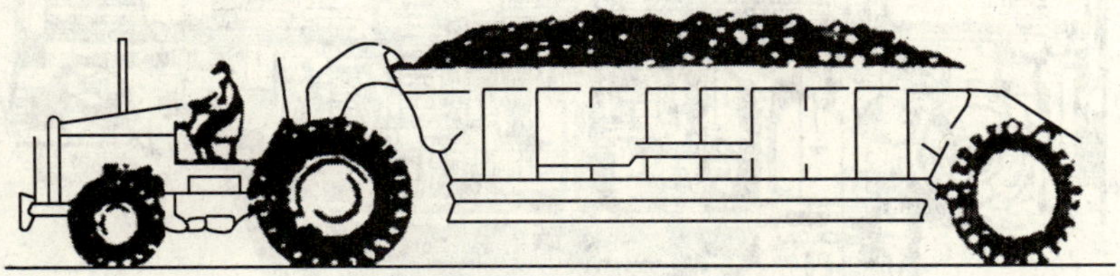

Fig. 3-22 Bottom-dump tractor-trailer unit with either drop or clamshell doors

3. Engines

Almost all large haulage units utilize the diesel engine, in either the four- or two-cycle version. Both types are available in horsepower ratings adequate for large haulage units. Turbochargers and intercoolers are required to achieve maximum horsepower output. A few gas turbines are in use on electric-drive units.

4. Power Trains

(1) Manual transmissions of the conventional gear box types are used on smaller trucks, usually with an air assist on the clutch. Separate retarder units for downhill operations must be provided as the manual transmission does not provide auxiliary braking effort.

(2) Power shift transmissions include torque converters and hydraulically clutch-operated planetary gear sets. Shifting can be either up or down under full engine power. It usually is automatic and depends upon converter output, turbine shaft speed and oil pressure. Automatic lockup into direct or overdrive is available. Hydraulic retarders can be incorporated ahead of the gear ratios and thus afford effective retardation in all speed ranges.

(3) Manual transmissions with torque converters employ a torque converter driving through an air-operated clutch and are used primarily where grades are relatively constant and few gear shifts are required. A dynamic retarder can be located ahead of the transmission.

(4) Electric-drive units utilize a generator driven by the engine to provide electric power to the drive wheels. The drive wheels can be driven by a conventional drive line and axle-powered by a d-c traction motor mounted ahead of the rear axle. Alternatively, the traction motors can be mounted outboard the frame and drive stub axle into the final drive gears. In the most popular electric-drive units, the traction motor is mounted integrally in the hub of the wheel to create the "wheel motor". The planetary gearing is arranged around the wheel motor. In both types of electric drives, excellent retarding is achieved through dynamic or electric resistance braking with the traction motors serving as generator units. Fig. 3-23 shows typical full power shift transmission incorpo-

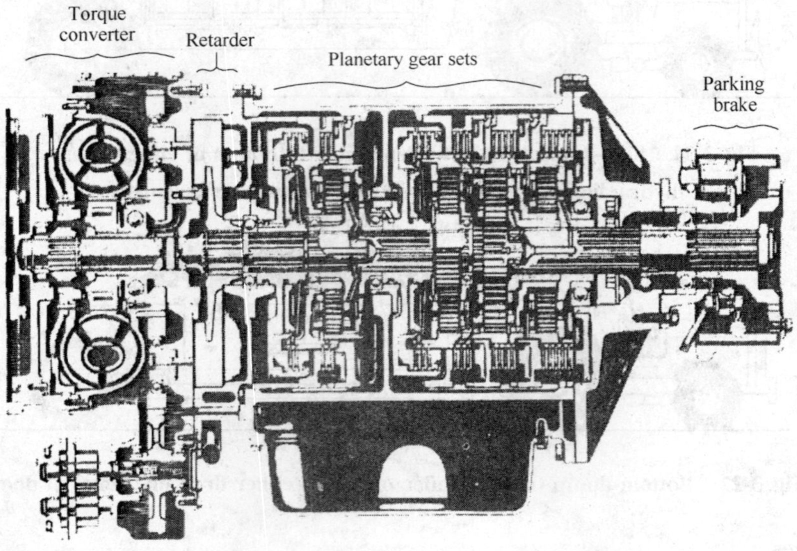

Fig. 3-23 Typical full power shift transmission incorporating torque converter, retarder and planetary gear sets

rating torque converter, retarder and planetary gear sets.

(5) Hydrostatic drive presently is used on small vehicles, but may have application on larger units in the future. In this type, the engine drives a high-capacity hydraulic pump, and oil is directed to hydraulic motors connected to each wheel. The direction of oil flow controls the rotation direction of the wheel.

Fig. 3-24 gives traction motor driving through a planetary gear reduction as applied in the wheel of an electric drive rear-dump haulage unit.

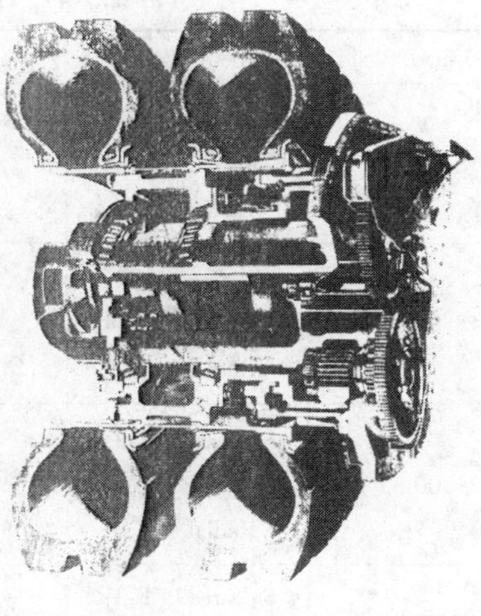

Fig. 3-24 Traction motor driving through a planetary gear reduction as applied in the wheel of an electric drive rear-dump haulage unit

5. Tires

Since tires are so important to the operating characteristics of the vehicle, and usually constitute the single most important cost element, it is imperative that consideration be given to the proper size, tread design, tread material, carcass design and material, etc. The Tire & Rim Association ratings usually govern the load that can be carried by the truck in terms of tonnage capacity. Recently, however, the use of a t · km/h factor has been developed to integrate speed of the unit into the tire loading. Since the rubber in the tire retains the heat generated by flexing under load while rolling, reversion from the vulcanizing process can take place at about 545℃. In addition, at temperatures below this point, the strength of the rubber is reduced and impact damage can occur. At the higher temperatures, the end result is usually ply separation and tire failure.

Heat generation in a specific tire depends upon:

(1) The weight the individual tire is carrying.

(2) The speed at which the tire is traveling.

(3) The ambient temperature of the surrounding air.

The t · km per hour (tkph) load can be calculated as follows:

$$\text{Tkph} = \text{Mean tire load} \times \text{avg. workday speed} \tag{3-2}$$

where

Mean tire load = [Empty weight on tire (t) + loaded weight on tire(t)]/2.

Avg. workday speed = Cycle distance (km) × trips per workday/total workday hr.

After the t · km per hour load has been determined, a tire can be selected with slightly great tkph rating than required. Since changes in materials and manufacturing methods occur frequently, specific tkph ratings should be requested from the manufacturer before the tire is purchased. Please consult tire supplier for current tire-design and tkph rating before purchase. Table 3-4 is tkph rating at ambient temperature, ℃.

Table 3-4 Tkph rating at ambient temperature, ℃

Tire size(U. S. std)	Type(U. S. std)	16°	27°	38°	49°
18.00-25	Std. tread (E-3)	223	207	185	164
	Extra tread (E-4)	178	164	148	133
	Radial S. C.	311	290	259	237
18.00-33	Std. tread (E-3)	259	230	207	178
	Extra tread (E-4)	223	207	185	164
	Radial S. C.	445	400	357	304
21.00-35	Std. tread (E-3)	296	274	245	214
	Extra tread (E-4)	252	223	209	171
	Radial S. C.	490	445	400	348
26.5-29	Std. tread (E-3)	259	230	207	178
	Radial S. C.	504	459	416	364
29.5-29	Std. tread (E-3)	274	245	223	193
	Radial S. C.	593	533	474	434
29.5-35	Std. tread (E-3)	319	297	267	238
	Radial S. C	785	696	549	416
33.25-35	Std. tread (E-3)	333	311	281	252
	Radial S. C	785	711	638	564
33.5-39	Std. tread (E-3)	362	319	281	237
	Radial S. C.	846	757	660	564
37.5-39	Std. tread (E-3)	407	371	341	304
	Radial S. C.	860	771	675	578
37.5-51	Std. tread (E-3)	430	385	341	290
	Radial S. C	1024	919	816	711
36.00-51	Std. tread (E-3)	742	652	578	519

3.2.2.2 Future Trends

The decade from 1960 to 1970 saw the development of open pit haulage trucks from small tandem-axle 37t units to large single-axle 203t units. Presently under development are haulage units of 254t. Engines have increased in size from 287kW diesel to 1,193kW diesel and 1,491kW gas-turbine units. Tires now in production use are of such size that a 1960-era truck would not be capable of hauling a set to the tire shop. However, it is doubtful if the next 10 yr will see such a continued increase in sheer truck size. Present trucks may be considered to be a product of evolution since, other than the development of electric-drive units, they are basically the same type of haulage unit being operated in the same manner. New techniques of a revolutionary nature will be required to attain the maximum productive capacity possible to insure economic pit operation. Trolley-assisted power systems for steep grades are becoming operational, paved roads are beginning to come into use, and electronic identification of both haulage units and type and amount of material being carried is almost a reality.

When unit and product data can be collected and transmitted remotely without reliance on expensive and delaying human observers, then total computerized truck dispatching will become practical. The economic benefits of a total electronic system will consist not only of more effective utilization of expensive haulage units, but a well-designed data collection system integrated into the dispatching computer program could approach a true on-line information system. It is conceivable that the mine superintendent will be alerted at 11:00 am by a recorded voice from the computer system that "Shovel 11 is running behind the required output of 0.56% ore and, unless corrective action is taken, will be 1,168t short by the end of the shift". Such information—acquired, analyzed and presented to operating managers while it is information, and not a day later when it is history—will be the real improvement in open pit truck usage in the decade of the 70s.

Professional Terms

non-ferrous metal mine 有色金属矿山
underground mine 地下矿山
underground haulage 地下运输
mine car 矿车
mine track 矿山轨道
rail-haulage system 有轨运输系统
locomotive 机车
truck 卡车
ore haulage 矿石运输
supply haulage 材料运输
transportation of men 人员运输
underground dumps 地下自卸设备
rubber-tired haulage 胶轮车运输
open-pit mine, surface mine 露天矿
surface haulage 地面运输
railroad haulage 铁路运输
truck haulage 汽车运输

Exercises

3-1 有哪些地下自卸设备?
3-2 请写出地下人员运输成本比较公式。
3-3 地面运输中使用的后卸汽车轮胎的发热量取决于哪些因素?
3-4 tkph 代表什么? 怎样计算?
3-5 MALT 代表什么?

4 Ventilation of Non-ferrous Metal Mines

本章介绍有色金属矿山的通风系统，包括井下大气特性、机械通风、自然通风、局扇、环境对有色金属矿山通风的要求、粉尘特点、粉尘取样设备及粉尘控制技术等。

4.1 Mine Atmospheres and Gases

4.1.1 Ventilation's Role

Importance of Ventilation and Environmental Control—Mine ventilation is necessary for the preservation of human life and for the conduct of operations involved in the underground extraction of minerals. It clearly is the most essential of the functions attendant to mining, yet it is auxiliary to the primary objective in mineral extraction. Ventilation in mining thus poses a paradox: ideally, it should be carried out with minimum interference with and cost to production operations, yet it is absolutely vital in providing adequate environmental conditions to support mining.

Deterioration of the atmosphere, along with rising ground pressure, imposes the ultimate constraint with depth in underground mining. Ventilation requirements and costs rise rapidly with depth. To preserve atmospheric quality under intense heat conditions, ventilation at great depths must be supplemented by air conditioning. Likewise, when mining at any depth under severe or excessive conditions of gas, dust or humidity, supplementary control measures must be employed.

In recent years, environmental standards in mines have been raised substantially. While limits are based on human tolerance, increasing concern is being expressed for standards of human comfort. The reasons are economic, as well as humanitarian. Worker performance and longevity correlates closely with environmental quality. Further, adverse accident and insurance compensation rates are a consequence of unsatisfactory environmental conditions. No mining company can afford to be lax in its environmental and ventilation practices.

Ventilation is the control of air movement, its amount and its direction. As a means of quantity control, it is one of the constituent processes of total air conditioning, the simultaneous control within prescribed limits of the quality, quantity and temperature-humidity of the air. Ventilation is the most common and most important conditioning process employed in mines for air environmental control.

In mine ventilation, air is coursed through the mine workings and openings themselves and, for auxiliary purposes, through vent tubing and ducts. Thus, many interdependencies exist between the mining system and ventilation method. In fact, the two often have evolved jointly, as in room-and-pillar extraction of bedded or tabular mineral deposits, such as coal. Ventilation in mines today is provided almost exclusively by large fans, although air flow may result due to natural (thermodynamic) conditions.

As mines increase in depth, size and complexity, demands on the ventilation system to maintain more stringent standards of environmental quality likewise rise. On the other hand, never have the technology and resources of the mining industry to cope with environmental problems been so well developed as they are today. In spite of rising costs, it is unlikely that further progress in mine ventilation and air conditioning must await technological breakthroughs.

4.1.2 Nature of the Mine Atmosphere

4.1.2.1 Air Mixtures

Air is a gaseous mixture comprising the natural atmosphere at the surface of the earth. Thermodynamically, it may be thought of as a mechanical mixture of dry air and water vapor (from 0.1% to 3% by volume). Table 4-1 shows the chemical composition of so-called pure dry air at sea level.

Table 4-1 Chemical composition of so-called pure dry air at sea level

Name	By volume, %	By weight, %
Nitrogen	78.09	75.530
Oxygen	20.95	23.140
Carbon dioxide	0.03	0.046
Argon, other rare gases	0.93	1.284

For ventilation calculations, the volume basis customarily is employed, and the composition is taken as oxygen 21% and nitrogen and inerts 79% (except when carbon dioxide is a contaminant). For air conditioning, the customary basis for calculation is a "pound of dry air", the only property remaining constant during psychrometric processes.

In normal atmospheres, dry air does not exist; it is always moist air. Further, there is no such thing as standard air, even though environmental "standards" exist. However, in mine ventilation work, air is assumed to be an ideal gas, to which the so-called simple gas laws apply.

4.1.2.2 Airborne Contaminants

Maintaining the chemical purity of the air within desired limits is termed quality control. It is accomplished in the mine atmosphere primarily through ventilation. Control of air environmental quality is more difficult in mines than in other industries because of the distance the air usually must travel, the contamination that may occur before the air reaches the working places, and the extent of the impurities generated in the working places.

Air for human breathing is, of course, the principal requirement of ventilation in mining. Various chemical processes also may consume oxygen and emit other gases, and mechanical and electrical processes require air for cooling just as do humans. Surprisingly little air is required for breathing; about $6m^3/min$ of fresh air containing 21% oxygen is adequate to maintain standards of 20% oxygen and 0.5% carbon dioxide.

Any undesirable substance not normally present in air or present in excessive amount is termed a contaminant. Vitiation of the air may result if threshold-limit values of environmental quality are exceeded (as prescribed for a working shift by a regulatory body, such as the U. S. Bureau of Mines or the American Conference of Governmental Industrial Hygienists). Although other contaminants occasionally occur, those most common in mine atmospheres are gases (nonparticulate) and dusts (particulate matter). The prime objective of quality control in mine ventilation is to contend with these two airborne contaminants, diluting and removing them.

4.1.3 Mine Gases

4.1.3.1 Nature and Sources

Although a wide variety of gases may occur in mines, only a few commonly are encountered as contaminants (see Table 4-2).

Table 4-2 Some gas contaminants in mines

Name	Effect	Source	Threshold limit value, %
Carbon dioxide	Asphyxiating	Combustion, strata	0.5
Carbon monoxide	Toxic, explosive	Combustion, IC engine	0.01
Hydrogen sulfide	Toxic, explosive	Strata	0.002
Methane	Explosive (5%—15%)	Strata	1.0
Nitrogen	Asphyxiating	Strata	80.0
Nitrogen oxides	Toxic	Combustion, IC engine	0.0005
Radon	Radioactive	Strata	1.3×10^5 Mev

4.1.3.2 Detection

The properties of some gases aid in detecting them, e. g., the rotten-egg odor of hydrogen sulfide or the reddish-brown color of the nitrogen oxides. However, only detection devices approved by the U. S. Bureau of Mines should be relied upon for detecting gaseous contaminants and for measuring their concentrations. Because of the highly explosive nature of methane, monitoring devices which continuously sample the air, sound an alarm and disconnect electrical power are employed as a safeguard in coal mines.

4.1.3.3 Sources

The most prolific sources of gases underground are the following:

(1) Natural—strata gas formed or trapped in the mineral deposit released when mine workings at atmospheric pressure are excavated. Methane in coal seams is most common. Release rate fluctuates with barometric pressure, exposed area of workings, mining rate and air flow.

(2) Man-made—gases generally produced by combustion processes, such as blasting, internal-combustion engines, fires or explosions, or by chemical action, such as human breathing.

4.1.3.4 Gas Control

Control of any atmospheric contaminant, gas or dust is based on these principles:

(1) Prevention (= avoidance).

(2) Removal (= elimination).

(3) Suppression (= treatment).

(4) Containment (= enclosure).

(5) Dilution (= removal).

They are listed in their preferred order of application, with prevention most desirable, although dilution in the ventilation airstream is most commonly and effectively employed. All principles may not be applicable in a specific case, and those listed first may not afford a viable choice at all. Ventilation alone, however, may be ineffective or prohibitively expensive in controlling airborne contaminants, and other measures should be employed prior to and in conjunction with ventilation.

A summary compilation of the quality-control measures most applicable to particular gaseous conditions follows.

(1) Control of strata gases—borehole drainage in advance of mining, bleeder entries around production areas, sealing off old workings, brattice or fan and tubing for auxiliary ventilation central ventilation.

(2) Control of blasting gases—reduction through good practice and proper selection of explosives, fume removal through exhaust system, localized and off-shift blasting, central ventilation.

(3) Control of IC engine exhaust—reduction through proper engine maintenance, high air : fuel ratio and proper choice of fuels, absorption and dilution of effluent in exhaust scrubbers, central ventilation.

(4) Control of fumes and gases from fires—isolation of fire zone by bulkheads, flooding or suffocation of fire, keeping fire zone under positive air pressure to reduce fume escape, dilution by central ventilation.

4.2 Mechanical Ventilation

The application of fans to induce airflow is common in the mining industry. The majority of modern mine fans are the axial-flow type. However, many centrifugal fans still are in use and may be preferred for some applications. The general fan laws are the same for either axial-flow or centrifugal fans. The only differences are in the individual characteristics of power, pressure and air volumes.

4.2.1 Fan Laws

These are as follows:

(1) Air quantity varies directly as fan speed. Quantity is independent of air density (twice the volume requires twice the speed).

(2) Pressures induced vary directly as fan speed squared, and directly as density (twice the volume develops four times the pressure).

(3) The fan power-input varies directly as the fan speed cubed and directly as the air density (twice the volume requires eight times the power).

(4) The mechanical efficiency of the fan is independent of fan speed and air density.

4.2.2 Selection of Fans

The performance of a fan in a ventilating system is determined by its characteristic curve and the mine resistance if acting alone on the system, and by its characteristic curve, characteristics of other pressure sources, and the resistance of individual zones of influence if acting in combination with other pressure sources. Fan characteristics are a matter of design, which is controlled by the manufacturer. The resistance of the mine is a matter of layout and maintenance of the ventilating network, and is controlled by the mine operator.

4.3 Natural Ventilation

Natural ventilation, or natural draft, is the term used to describe mine airflow resulting from a pressure differential caused by unequal densities or weights of two connected air columns. The unequal densities, although influenced by other factors, are due mainly to the transfer of heat or thermal energy from the underground rock strata to the mine atmosphere.

With the exception of very shallow mines underground air temperatures approach strata temperature and remain fairly constant. Consequently, temperature, density and total weight of the upcast air column will not fluctuate to any greater extent. The density and total weight of downcast air column, however, is directly influenced by and fluctuates with surface-air temperature. Since a density change in either of the columns will cause a change in the natural-draft pressure, this pressure differential may vary from a maximum in one direction to no differential to maximum differential in the opposite direction, depending on whether surface temperature is lower than, equal to or greater than underground temperature. This results in unpredictable volume flow, which may vary from a maximum in one direction to flow in the opposite direction. In some instances, natural ventilation may maintain underground air-quality standards but generally it is not a dependable means for continuous and effective ventilation. To this extent, U. S. regulations under the Coal Mine Health & Safety Act of 1969 require mechanical ventilation in all underground coal mines.

4.4 Booster Fans

The development of metal mines, because of irregularities in the ore body, often results in air splits or isolated mining areas which offer excessive high resistance to airflow. Where it is impractical or uneconomical to induce flow by decreasing the resistance or regulating the low-resistance split, booster fans may be installed in the high-resistance split. The energy generated by the booster fan is used to overcome the resistance of the split in which it is installed.

Booster fans may be installed at any location in the split but, to minimize leakage, the fans should be on the same side of the circuit as the main fan—that is, on the return side with an exhaust fan and on the intake side with pressure fan, and as near the junction of the main split as practical.

If two openings are available, the fan may be installed in a bulkhead on one of the openings, with airlocks in the second opening to separate the intake and discharge sides of the fan. Where only one opening is available, a runaround may be driven or an offset in the entry may be excavated for fan installation. Fig. 4-1 illustrates the latter. All such installations should be of fireproof construction.

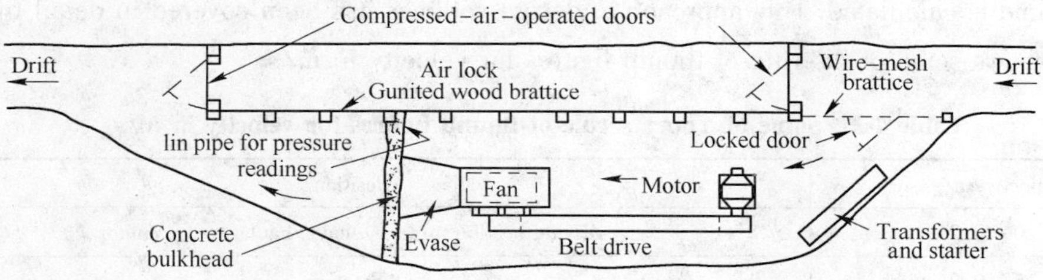

Fig. 4-1 Typical underground installation of fan in booster position

Determining the duty of a booster fan may appear difficult. However, if its position in the ventilation circuit is evaluated, a suitable selection can be made. If the fan is located on a minor split, the desired volume usually is known and the pressure required can be calculated disregarding any effect on main fan operation. This does not apply to fans operating in series with the main or to fans operating on a major split which handles a large part of the total quantity. Duty of fans in this position is determined by calculation or graphic solution.

4.5 Primary Ventilation Systems——Non-ferrous Metal Mines

Mine ventilating systems vary greatly and are dependent upon the size and shape of the ore body, system of mining and depth of operations. The most difficult systems to establish and maintain are in deep multilevel operations mining irregular ore bodies. The resistance to airflow of these systems requires the efficient maximum utilization of air volumes necessary to provide and maintain a safe and healthful underground atmosphere. This can be achieved only by the proper distribution and control of adequate air volumes.

There is no ideal or standard system of mine ventilation. The effectiveness and efficiency of the system will be determined by how well certain basic fundamentals are applied and maintained.

4.5.1 Air Volumes

Two factors—adequate volumes and proper distribution—along with a working knowledge of the mining system, from the basis for efficient mine ventilation. Unfortunately, there is no simple method of determining adequate volumes. The quantity and physiological effects of all contaminants introduced into the system must be evaluated.

Volume requirements for a projected mine sometimes are based on the volume circulated through an active well-ventilated mine of similar tonnage. With this approach, a complete analysis and comparison of mining methods, mining equipment, contaminant-control measures and air distribution should be made between the proposed and active mines. A variance in any one of these factors could lead to under- or over-design of the proposed system unless appropriate adjustments are made. Complete reliance should not be placed on this approach. However guidelines in certain areas of volume comparisons are very helpful in system design.

Other ventilation experts use a minimum velocity in various active mining areas from which design volume is calculated. This approach to design volumes has been covered in detail by Yourt. Table 4-3 shows some of his rule-of-thumb figures for velocity in m/s.

Table 4-3 Some of Yourt's rule-of-thumb figures for velocity in m/s

Velocity, m/s	Position
0.20—0.25	Driving headings, raises, shafts, haulageways, chutes
0.15—0.25	Stopes
0.25—0.88	Scrams
0.77—1.02	Enclosures, such as dump doors and crushers
0.25—1.02	Loading pockets

These figures, no doubt, are based on a source of uncontaminated air. Additional volume for dilution will be necessary where contaminated air is involved.

One area where definite volumes may be designated involves the use of diesel equipment. Regulating agencies usually specify conditions under which such equipment can be used. The code of the Saskatchewan Dept. of Mineral Resources, specifies the requirements which diesel exhaust gases and the atmospheres adjacent to the engine must meet for underground operation. Analysis of exhaust gases and conditions of operation enable the engineer to calculate the required air volumes. Approval plates issued by the U. S. Bureau of Mines specify the volume of air necessary for each piece of diesel equipment tested and approved for underground use.

Sampling and subsequent analysis is the one certain method of determining the quality of the working atmosphere, from which the volumes necessary for proper quality control can be determined.

4.5.2 Method of Mechanical Ventilation

Underground mines can be ventilated by the force (overpressure) or exhaust (underpressure) method. There is no simple rule to follow in making the selection. Management must weigh the factors involved and choose the method best suited for anticipated conditions. Considering identical ventilation circuits, each adapted for the method employed, the variation in efficiency between force and exhaust systems is negligible.

With the force system, the main shaft usually is on return air, which often is detrimental to health and safety and may have a harmful effect on hoisting equipment and shaft installation. On the other hand, downcast air in winter may result in severe icing, accompanied by increased safety hazards and possible production delays. Freezing in the downcast hoisting shaft has been overcome by introducing heated air below the shaft collar.

The advantages and disadvantages must be carefully weighed, but in general the exhaust system is preferred with the hoisting or main shaft on intake air.

In shallow uranium mines with fractured ground, or in areas of contiguous mining where ground cracks into abandoned mines are possible, the force system may be the only practical method.

4.6 Ventilation for Environmental Control——Non-ferrous Metal Mines

4.6.1 Health and Nuisance Considerations

Fine particles of dust are a byproduct of the various operations associated with the extraction of ores and the winning of metals. They constitute both a health hazard and a nuisance. The latter condition arises when the dust is dispersed in the form of clouds capable of reducing visibility, adversely affecting morale, and causing undue wear and premature failure of components in mechanical equipment and sensitive electronic systems. A direct result is an increase in mining costs due to increased accident frequency, undue delays while waiting for dust and fumes to clear, and excessive maintenance and repair costs.

The health hazard involves the inhalation of fine dust particles and their retention in the alveoli of the lungs. The degree of the hazard will depend upon exposure time and nature of the dust, particularly its concentration and physicochemical characteristics. Pneumoconiosis, a term used to describe all lung diseases caused by dust, is the result of overexposure, and an incidence rate of 1 per 100 underground workers annually is not uncommon in many mining districts.

To reduce the incidence of pneumoconiosis it is necessary to provide a relatively dust-free working atmosphere—that is, one where the quality of air meets the standards laid down by industrial hygienists. A useful guide for this purpose is one published annually by the Committee on Threshold Limits of the American Conference of Governmental Industrial Hygienists. These values are reviewed annually by the committee and revised as conditions warrant. However, it should be noted that these maximum allowable concentrations and threshold-limit values are intended only as guides and should be interpreted and applied only by persons trained in the field of industrial hygiene. The proposed limits for quartz dust are the subject of review and evaluation in a publication by Ayer.

This section reviews the characteristics and the sources of airborne dust particles associated with underground mining operations and, in a general manner, discusses some of the common techniques for control of dusts, such as suppression, dilution and removal. The reader is referred to an informative text prepared by the International Labor Office for a comprehensive coverage of the use of techniques other than ventilation for the prevention and suppression of dust at all mining operations.

4.6.2 Dust Characteristics

The design and implementation of effective dust-control measures can be facilitated by a good working knowledge of the behavior of airborne dust particles. For the purpose of this discussion, airborne dust is particles having a diameter less than 10μ. This classification includes those particles in the respiratory range (0.25 to 7μ) which cause pulmonary damage.

In a discussion of fine-particle dynamics, Walli presents some practical examples of dust-particle trajectories and settling velocities. Using the equation for calculating the terminal, or settling, velocity for dust in the turbulent region for airflow, he shows that a particle 10μ in diameter with a specific gravity of 2.7 will gravitate at a speed of 0.00762m/s. In practice, this velocity is insignificant when compared to stray currents of 0.00762 to 0.1016m/s created by a person walking in a still atmosphere. Furthermore, if the same dust particle is given a velocity of 13.4m/s, it will travel less than 305mm in a horizontal direction before terminal velocity is reached and it begins to fall vertically at a rate of 0.00762m/s. Consequently, it is apparent that fine dust particles do not retain appreciable control over their settlement for more than a fraction of a second and, for all practical purposes, that they behave in a manner similar to the air which encompasses them.

This discussion leads to some conclusions of great importance in dust-control strategy. One is that the engineering problem of dust control is resolved to designing a system for controlling the air involved in the dust-producing process. Another important conclusion is that dust in the respiratory range will remain in suspension indefinitely so that, in the confined passages of an underground operation, dust clouds can readily be transported to contaminate areas far removed from their source. It is apparent that every effort should be made to prevent dust from becoming airborne.

For a comprehensive coverage of particle dynamics, the reader is referred to publications by Hemeon, Dalla Valle and Lapple et al.

4.6.3 Assessment of Air Conditions

One of the most important features of an effective program for minimizing exposure to dust at underground operations is accurate assessment of environmental quality. The adoption of a schedule for routine surveys of dust concentrations will pinpoint the locations and the severity of dust-producing operations. Furthermore, the surveys will determine whether the measures deigned for dust-control purposes are effective, and whether they are being utilized and properly maintained.

4.6.4 Survey Procedure

The instruments used in making ventilation surveys, together with the procedures recommended for monitoring air conditions, are outlined in detail in publications by McElroy, Kingery, Yourt, and the ILO. In general, the method should comprise a comprehensive survey of all ventilating currents and representative sampling of airborne dust at all operations. Relevant data on the performance of fans, the volume and direction of all airflows, and the concentrations and locations of dust samples taken should be plotted on mine plans and sections, which should be kept up to date. It is suggested that other information, such as the psychrometric condition of the airflows, and the location of fire hydrants, hoses, extinguishers, doors, stopings, etc., also should be recorded on the plans. (Information of this nature will be invaluable in the event of a mine fire.) The finished mine plans will provide a good perspective of the existing air conditions and serve as a useful guide in determining the type and location of remedial measures necessary to improve and maintain the quality of the environment.

4.6.5 Dust-sampling Instruments

The lack of conclusive data concerning the pathogenesis of the pneumoconioses renders it difficult to state, with any degree of certainty, the dust parameters that should be measured in evaluating the health hazard. It is beyond the scope of this presentation to enter into a technical description and discussion of the numerous instruments available for dust-sampling purposes. Many are amply described and illustrated in the text, "Air Sampling Instruments". The popular trend on this continent, and to some degree in Great Britain and Europe, is toward the gravimetric method or, as the name implies, collecting a dust sample for assessment by mass or weight. Recent developments in this field have been directed toward the collection of only the respirable portion of the dust cloud, utilizing instrumentation such as that described by Ayer and Yourt. Included in the desirable features of this method are acquisition of an accurate estimate of the respirable portion of the dust cloud and an adequate amount of sample for analytical purposes (for example, analysis of quartz content by X-ray spectroscopy).

Another recent development in this field is the "Diffraction Size-Frequency Analyser", described by Talbot. This method utilizes the phenomenon of light diffraction to measure the total and respirable surface areas of dust particles collected on glass slides by a thermal precipitator. One disadvantage of this method is that the sampling period is limited to less than 1 hr.

It would appear that the ideal instrument for dust assessment should be capable of fulfilling the needs listed in a paper by Bardswick and White:

(1) Sampling for a period of at least one working shift.

(2) Collecting airborne dust particles representative of the dust cloud in the worker's breathing zone.

(3) Collecting an adequate sample of dust for chemical/mineralogical analysis.

(4) Determining the number and size distribution of the respirable particles constituting the dust cloud.

(5) Determining the surface area of the respirable dust particles as a means of assessing their biological activity.

(6) Determining the mass of the respirable dust that may be inhaled, and possibly retained, during the working period.

The ultimate in the design of a dust-assessing instrument would be an arrangement that would simulate the human lung and preferably have the same retentive capacity as shown on a graph depicting the alveolar retention of particles in the normal respiratory system. Since the ideal type of instrument is not available, it is suggested that the ventilation engineer consult the state and federal industrial-hygiene codes for guidance in the choice of appropriate dust instrumentation and survey techniques.

4.6.6 Dust-control Techniques

The exposure of workers to harmful dust concentrations may be reduced by a systematic approach to the problem which would include implementing all or some of the following:

(1) Reducing the production of dust.

(2) Preventing the dispersal of dust clouds.

(3) Providing dilution ventilation.

(4) Utilizing personal protective measures.

4.6.6.1 Dust Production

Although most underground operations (such as drilling, blasting, conveying, etc.) produce copious amounts of dust, the amount so produced can be reduced in some instances. The environmental engineer should make a critical survey of all equipment and processes with a view to improving the use of machinery and altering techniques to produce less dust. It is an established fact that dull drill bits, through their grinding action, produce more dust than sharp ones, and that insufficient delivery of water to the cutting edge to remove rock debris also results in larger amounts of dust. In other instances, the grinding or crushing action of falling rock may be reduced by decreasing the distance through which the material falls—for example, a belt-transfer point or a chute used for loading ore trains or feeding a rock crusher. These are but a few examples of the many ways in which the production of dust can be reduced and, needless to say, any success in this approach will greatly enhance the success of the overall dust-control program. In practice, the dust-laden air should be exhausted directly to a return airway or rendered clean before reuse.

4.6.6.2 Dust Dispersal

The dispersal of dust can be reduced and controlled by a twofold method involving the use of water and control of air currents. Although water and other wetting agents are not significantly effective in allaying airborne dust, they are very useful in suppressing dust at its source. For this reason, every effort should be made to apply the wetting agent directly at the source to ensure that the broken rock is wet in situ and conditioned for ensuing processes tending to disperse dust.

The dispersal of dust can be practically eliminated in most instances by confining the dust-producing operation within an enclosure and controlling the air contained therein. The vitiated air from within the enclosure can be exhausted directly to the upcast airway or, if this is not feasible, it can be delivered to a filtering plant where the dust concentration can be reduced to an acceptable level.

4.6.6.3 Dilution Ventilation

The use of ventilating currents to dilute and remove dust clouds is the most common method of controlling contaminants at underground operations. This procedure is carried out in producing areas, such as stopes, scraper drifts, etc., by directing an air split from the main ventilating stream through the workings. The design criterion for volume of air required is based on an air velocity of 0.15 to 0.25m/s, depending upon the type of operation and other local conditions. In some instances, volume may have to be increased greatly—for example, high-speed drives or scraper drifts, where the severe dust-producing operations may require as much as 0.76m/s or more.

In headings and raises, the design volume also is based on providing an air velocity of 0.15 to 0.25m/s or 0.0142 to 0.0236m^3/s per sq m of face. The type of auxiliary ventilating system employed will be governed by the local conditions at the mine. In most cases, the push-pull (or overlap system) will provide the most satisfactory environment. This system consists of a main ventilating duct which exhausts vitiated air, and a small blowing line kept to within 6 or 9m of the face. The length of the overlap of the exhaust and blowing lines will depend upon the size of the drift and, in any case, should be not less than 9m. The blowing fan should be rated for the calculated volume and, to avoid recirculation of vitiated air to the face, this figure should not exceed approximately 60% of the exhaust volume.

The overlap system should not be adopted for those working areas subject to a high virgin-rock temperature. Under these conditions, the cool fresh air is heated and becomes saturated with moisture during the time it takes to reach the working face. To avoid this, the forcing or blowing system of auxiliary ventilation should be installed so that the fresh air can be kept dry and delivered to the face with a reserve of cooling power. Ideal conditions will prevail if the discharge end of the blowing system is kept to within 6 to 9m of the working face. The sweeping or flushing action can be improved by installing a cone-shaped duct on the discharge end to increase the outlet velocity, or "throw", of the ventilating current.

Blasting damage to the ventilating duct may be avoided by utilizing a semirigid polyethylene duct with a wall thickness of 6 to 10mm. St. Pierre reports that the performance of this type of duct is excellent, with minimum leakage at the joints, a very low friction factor and 100% salvage value. He also reports that conditions were greatly improved during drilling operations in raises by installing a 152mm-diam polyethylene duct for ventilation purposes. With the discharge end of the duct kept to within 7.62m of the face, dust concentrations were reduced to well below acceptable limits. Huggins reports that dust concentrations in raises were reduced approximately 50% by use of a compressed air venturi and a 76mm-diam aluminum pipe for ventilation purposes.

The use of exhaust systems only for auxiliary ventilation should be discouraged, since it is impractical to keep the end of the duct near the dust-producing operation, particularly when blasting. The effectiveness of an exhaust duct diminishes rapidly and approaches zero at a distance of approximately two pipe diameters from the inlet end.

4.6.6.4 Personal Protection

While the use of personal protective measures generally is considered a last resort in a dust-control program, there are occasions when their application can be very effective. One of the indirect methods used in all mines is removal of workers from the area at the time of the blast, and for a sufficient time thereafter for complete removal of the contaminants. A very commendable practice is to schedule all primary blasts, and as many of the secondary ones as possible, for the end of the working shift. The exposure to dust and fumes thereby is considerably reduced because the workers will be out of the zones through which the vitiated air travels on its way to the main upcast airway.

There are occasions when local conditions make it impractical or impossible to confine effectively a severe dust-producing operation. Under these conditions, personal protection may take the form of an enclosure large enough to house the worker and maintained under positive pressure of good-quality air. If this is not feasible, the last resort is use of a dust respirator approved for that particular type of dust by the U. S. Bureau of Mines. However, it is suggested that every effort should be made to utilize some other means of dust control, since experience has shown that the use of dust respirators is difficult to promote, either by a voluntary or a mandatory approach.

4.6.6.5 Drilling Operations

Drilling is an essential operation in metal mines and the percussive type of drill in common us can be the source of copious amounts of dust. However, a number of precautions can be taken to improve this undesirable feature.

The "collaring" operation produces considerable dust, particularly in the absence of water. Introduction of the automatic back head (interlocking air and water valves) was designed to eliminate the temptation to "collar" dry. Unfortunately, some operators install a hand valve in the water hose near the drill so that dry collaring still is possible—and practiced in many mines. These hand valves should be removed or relocated so that they are not readily accessible to the driller. Dry collaring also can result from careless operation of the automatic control. As the latter is advanced to the "On" position it opens the water valve before the air throttle. If the operator "slams" the automatic control to the open position, the drill steel begins to rotate before any water reaches the cutting edge. This bad practice should be eliminated by instructing the operators in the proper drill operation.

Percussive drills continue to liberate dust after the collaring operation is completed, even though the drill hole is full of water. This is caused by the leakage of small amounts of compressed air past the piston into the front head of the machine, whence it travels down the hollow drill steel as bubbles in the water. At the toe of the hole, these air bubbles entrain dust particles, which are released as the bubbles emerge from the hole. Drill manufacturers have managed to overcome this problem by designing the vented front-head machine, in which the compressed air is permitted to escape before it enters the shank of the drill steel. Although this type of machine disperses less dust than earlier models, it is not used extensively in North America because of the high maintenance cost of the parts in the drill front head. In the standard machine, leakage of compressed air past the cylinder is intentional, since this air is saturated with oil which lubricates the moving parts in the front head of the unit. It would appear that the vented type of drill will not be fully acceptable until the manufacturers implement an alternative method of lubricating the front-head parts of the machines.

A number of other features can adversely affect the operation of drills, including the size and condition of the water tube, the sharpness of bits, the condition of the water hole in the bit and the shank end of the drill steel, and the operating water pressure. These and some other aspects in drill operation, with recommendations to reduce the amount of dust produced and dispersed, are reported by Yourt and Bloomer.

4.6.6.6 Blasting

Blasting is unlike most other mining operations, since it produces not only dust but other contaminants, such as carbon monoxide and nitrous fumes. Since little can be done to prevent the production of dust during a blast, the emphasis should be on means to control the resultant contaminants.

The first step in controlling dust produced by blasting is to ensure that the area surrounding the blast (the walls, floor and back) is thoroughly wetted beforehand. This precaution will prevent dust settled out during previous operations from becoming airborne. Furthermore, some of the dust created by the blast will adhere to the wet surfaces in the area, thereby reducing the concentration in the air stream.

The second step, which is the most effective measure for controlling the contaminants resulting from a blast, is use of an arrangement called the "air-water blast". This device is excellent in minimizing the dispersal of dust and in reducing the amount of nitrous fumes (because of their solubility in water). Unfortunately, the carbon-monoxide content of the air from a blast is not appreciably affected since it is practically insoluble n water.

The design of the air-water blast device varies from a simple inexpensive model assembled in the machine shop at the mine to the elaborate expensive type of unit available from a number of manufacturers. The effectiveness of this device is determined by the fineness of the mist or fog produced, and the "throw" and "spread" of the mixture. An unpublished report, covering tests made on 11 different types, indicated that the water consumption varied from 0.06 to 0.61L/s; compressed air, 0. 0496 to 0. 1416m^3/s (free air). It is apparent that a particular design should not be adopted as standard for use in all working places in all mines. A working area with little or no ventilation would benefit most from the unit providing the high rate of airflow. On the other hand, it would be poor engineering to waste expensive compressed air in a place well ventilated by conventional means. Similar reasoning may be applied to the utilization of the water delivered, particularly for blasts in drifts and raises vs. the larger ones in stopes involving much greater tonnages of broken material.

An important feature of the air-water nozzle is the manner in which the compressed air and water are blended before dispersal. The design should incorporate means precluding the drainage of water into the compressed-air line, particularly when the pressure of the latter is lower than that in the former. Two suitable types of nozzle are illustrated in the text published by the ILO, and the details of a number of others are available on request from the Mines Accident Prevention Association of Ontario.

The final step in the control of contaminants resulting from blasting is providing means to cleanse pollutants. A unique type of collector used in South Africa consists of a bed of vermiculite, treated with a solution of sodium carbonate and potassium permanganate, which removes the nitrous fumes, and a flannel filter for the collection of dust particles. After treatment, the air is returned to a fresh-air split where the dilution ratio is at least 5 to 1. Plant details are presented by Rabson.

In the absence of a filtering system, blasting dust and fumes should be diluted and exhausted to surface via an untraveled route, preferably an upcast raise designed for that purpose. If this is not feasible, the blasting schedule should be arranged so that the vitiated air will pass through working places when the miners are absent. Failure to do this creates and undesirable health hazard and reduces production efficiency due to miners leaving their working places for the time necessary to clear the dust and fumes.

4. 6. 6. 7 Mechanical Loading

Utilization of an air-water blast for a suitable period after blasting obviates the need for applying a water spray immediately before commencing the loading operation. On the other hand, if the rock pile has to be wetted by means of a hand-held hose, the quality of the environment will deteriorate considerably because the application of water under these conditions is similar to directing a stream into a barrel of flour.

Investigation of a rock pile after it apparently is soaked with water will reveal that the depth and extent of water penetration are shallow and limited. In other words, it is practically impossible to wet a pile of rock thoroughly by conventional spraying methods. As a result, dry dusty material is exposed continuously as the loading or scraping operation progresses. Under these circumstances, it is necessary to maintain a continuous mist or water spray at the point where new fragmental material is exposed. Use of a hand-held hose is not recommended, because the operator, on exposing new rock during the process of loading an ore carrier, may be tempted to complete the cycle before applying more water. Some success has been achieved by installing a nozzle spray on the front end of compressed-air-operated loading machines. A nozzle which provides a flat diverging stream of mist-size water droplets produces a good wetting action with minimum water consumption.

The scraping of rock in stopes and scram drifts produces large amounts of dust, particularly if the material is dry and the speed of the operation is high. A successful method of controlling this dust is to install a bar-type water sprays where the scraper "bites" into the rock pile, with one or more additional nozzle-type sprays at intervals over the path followed by the scraper. A mist type of spray at the ore-pass or loading point will reduce dust dispersal and condition the rock for ensuing operations (such as crushing and conveying). Details of the nozzle and bar-type sprays are presented by Peacock. The effectiveness of these units in reducing dust dispersal is also discussed.

4.6.6.8 Ore and Waste Passes

Ore and waste passes are, by the nature of the operation carried on therein, sources of large quantities of airborne dust. The broken rock delivered to the passes contains a considerable amount of inherent dust as a result of the comminution effects of preceding operations, such as blasting, loading, etc. Furthermore, the autogenous-grinding action of the rock as it is dumped and falls down the pass produces more dust, which becomes airborne and subject to dispersal. The extent of the dispersal of dust at these locations is governed by two factors: (1) the displacement of air in the raise by the rock, and (2) the entrainment of air due to the "piston" effect of the falling rock in the raise.

A number of measures can be employed to reduce the dispersal of dust at the entrance to ore and waste passes. The first line of defense is to ensure that the rock is thoroughly wetted before delivery to the dump. As previously mentioned, the use of water-spraying devices in preceding operations will tend to suppress the dust contained in the rock mass. A further wetting effect can be obtained at the dumping site by installing a mist-type atomizer to spray the rock as it falls into the pass. A very fine mist should be used to provide the optimum wetting effect with a minimum amount of water. Excessive use of water at the orepass can be objectionable for two reasons: (1) the excess water in the ore can adversely affect crushing and milling operations, and (2) a large quantity of water may accumulate on top of the ore in the raise, thereby creating a hazardous condition for workers on the lower levels.

The second step in providing a good environment at ore and waste passes is to prevent the escape and dispersal of dust into working areas by confining it within the passes. This can be accomplished by a system of stopings and airtight doors over the dumps or "tipping" points. The maintenance of these doors is of prime importance, since their main purpose is to prevent the escape of dust-laden air. A number of designs have been developed over the years and a typical one is described by Kneen. This particular unit was designed for Granby-type cars. It is pneumatically operated and easy to maintain.

The third step is providing means to keep the confined ore or waste pass under negative pressure to ensure that all leakage paths are indraft, and to capture the air displaced when rock enters the raise. This can be accomplished by installing a suitable fan to exhaust from a convenient point in the raise. The vitiated air exhausted must be filtered or sent via a direct untraveled route to the return-air raise.

The ideal location for the exhaust fan is at the top of the ore or waste pass. Under these circumstances, the coarse dust is allowed to settle out in the raise, thereby reducing the load on the dust collector and, in the absence of a cleaning device, minimizing the abrasive wear on fan blades. Another advantage of this arrangement is that it facilitates the disposal of dust-laden air by utilizing an exhaust point above the active or producing levels of the mine.

An alternative to the use of a single fan is to install smaller individual exhaust units to service each dumping area. Suitable enclosures, fitted with hinged or sliding doors, can be constructed over the dumping area. The rating of the fan is determined by the volume of air necessary to maintain an indraft velocity of 1m/s through the door opening during the dumping operation. The air exhausted from the raise usually is very dusty and must be delivered to the return-air system or processed in an efficient dust collector before joining a fresh-air split. The details of a particular design will depend upon the local conditions at the dump site. The reader is referred to typical layouts described by Kneen, Phimister and Gray et al.

Although the dust control measures previously discussed are capable of capturing the air displaced by the rock dumped into the raise, they are not entirely satisfactory for the control of dust dispersed by airflows induced within the raise by the "piston" effect of falling rock. A practical solution to this problem is described by Marshall. The success of this method is due to the provision of "relief valves", created by driving connecting drifts between the ore and waste passes at a number of elevations. In practice, the dust-laden air under pressure because of the falling rock is short-circuited through the connecting drift to the parallel raise, which acts as a "reservoir" for the air exhausted by the fan. In the case study by Marshall, he reports that the surges of dusty air were practically eliminated and that the success of the system enhanced dust-control effectiveness at the underground crusher station and loading pocket.

At first glance, it may appear that the problem of controlling the dispersal of dust from ore and waste passes has been solved. Unfortunately, this is not so, because there is no proven method of accurately determining the fan rating required to overcome the effects of the surges of dusty air. The volumes of the induced airflows depend upon a number of parameters, including: the size and flow rate of the falling rock, the distance through which it falls, the size of the raise, and the openings available for airflow into and out of the system of raises. Although this has been the subject of a number of research projects, the answer still is evasive and there exists a need for further investigation. The nearest approach to a solution has been reported by Anderson, who presents the formula

$$Q = 10A^3RS^2/D \tag{4-1}$$

where Q is the induced airflow in cubic feet per minute (cfm), A, the opening in square feet in the upstream enclosure, R, rate of material flow in tons per hour, S, height of fall in feet, and D, average particle diameter of material in feet. (Note: 1 cfm = 1 cu ft per min = 0. 000472m^3/s; 1 m = 3. 2808 ft; 1 t = 0. 9842 ton)

Although this formula has limitations and does not provide an accurate solution in all instances, it can be a very useful guide in eliminating some of the guesswork involved in determination of a suitable fan rating for the exhaust system.

Another method of controlling the dispersal of dust from an orepass is to exhaust the induced air from a point near the bottom of the raise. In this approach, the volume of air exhausted must be sufficient to maintain the raise under negative pressures, thereby precluding the buildup of positive pressures, which cause the surges of dusty air. This method can be used to good advantage in a multilevel mine where ore is continuously delivered to secondary orepasses which feed a main orepass (for example, a caving method of mining incorporating scraper drifts). The application of this method is discussed by Foster. One disadvantage is the problem created by the coarse dust particles carried out of the orepass by the air being exhausted. A plenum area or some other provision must be made for the removal and disposal of the coarse-dust particles. Another disadvantage is the abrasive effect of dust particles upon the blades of the exhaust fan, particularly if it is installed in the section of the return airway near the orepass.

4. 6. 6. 9 Crushers

An underground crushing station is a confined space that can be contaminated readily by dust if preventive means are lacking. The deterioration of the environment in these rooms is caused by the dust contained in the ore as a result of preceding operations, plus that produced by the grinding action of the crusher jaws. These sources of dust can be controlled by applying the usual principles (1) confinement and (2) dilution and removal of the particles which become airborne within the enclosures.

The simplest and most effective approach is to enclose completely the chute which feeds the crusher, the jaws of the crusher and all other openings to the path followed by the ore stream. To prevent the dispersal of dust, the enclosure is kept under negative air pressure by a fan which exhausts from the opening below the crusher jaws. The entry to the exhaust duct in the crusher pit should be as far away from the ore stream as practicable to avoid the capture of coarse particles which have a detrimental effect on all components of the dust-control system.

The rating of the exhaust fan is determined by calculating the volume necessary to provide an indraft velocity of 1 m/s through all unavoidable openings and leakage paths in the enclosures. Since the volume of air required is a function of the area of the openings available for airflow, it is apparent that the fan rating will increase considerably if the jaws of the crushers are not enclosed. Although some objections may be raised to total enclosures of feed chutes and crushers, there are numerous examples in the literature of the successful implementation of this design. Some typical layouts are described by Walker, Slater and Gray et al.

Disposal of the vitiated air exhausted from the enclosures may be accomplished in a number of ways, the choice of which will be governed by local conditions. The ideal method is to exhaust the dust-laden air through a ventilation duct leading to the return-air raise. Unfortunately, this is not always practicable because of the location of the crusher station relative to the raise. Consequently, an alternative method must be employed. In the alternate design, a suitable dust collector is installed in or near the crushing station to filter the air before it is returned to a fresh-air split. If the filtered air is to be recirculated within the crushing plant, it should be diluted by a ratio of three volumes of fresh air to one of filtered air to compensate for the inefficiency of the dust collector. The effectiveness of the latter will depend upon a number of factors, such as the dust loading, the type of filter medium, and the air velocity through the unit.

The type of dust collector recommended for use in underground dust-control systems is one which incorporates a fabric material as the filter medium. It provides good filtering efficiency at relatively low cost. The three types of fabric most commonly used are wool, terylene and cotton. For many years, the latter was very popular and it still is widely used in some of the larger asbestos plants. The main disadvantage of the cotton-type collector is the space requirement due to the low air-to-cloth ratio (velocity through the cloth) required for optimum filtration. If the air velocity through the cotton fabric is increased beyond 0.01m/s, the efficiency of filtration deteriorates rapidly.

The use of wool as a filter medium originated in South Africa and it still is used extensively there and in Australia. Permeability is low, resulting in good filtering efficiency at high air velocities. Rabson, Walker and Phimister report on underground installations where good air conditions were maintained with air velocities of up to 0.13m/s through flannel bags. However, the life of the bags was limited to about 20 mo in one installation and, in most instances, the filter media is removed from the housing for cleaning purposes. These two features augment the operating cost of flannel-type filters, and the inconvenience of the high maintenance requirement merits due consideration.

In recent years, the development of synthetic fabrics has introduced a number of new cloths suitable for air-filtration purposes. One of these, a heavily napped terylene, has gained much prominence in the Canadian mining industry. Segsworth reported on the use of this material at an air-to-cloth ratio of 7 : 1. After 7 yr in an underground crushing station, the original terylene filter bags still were in place and delivering clean air with dust concentrations well below the threshold-limit values. Similar performances are being obtained at other Canadian mines under conditions of high humidity, indicating that a heavily napped terylene can be recommended for most underground applications.

One feature of a dust-collecting system often overlooked is disposing of the collected dust. A poor method can result in dispersal of the dust in the working area, thereby negating the value of the control system. Proper disposal of dust will be greatly facilitated by installing the dust collector near the orepass so that the dust hopper can be drained directly to the raise via a duct or totally enclosed chute.

Professional Terms

ventilation 通风
airborne contaminant 空气污染物
mine atmosphere 井下大气
mechanical ventilation 机械通风
natural ventilation 自然通风
booster fan 局扇
mine ventilation system 矿井通风系统
air volume 风量
dust 粉尘
airborne dust 大气浮尘
carbon dioxide 二氧化碳
dust-sampling instruments 粉尘取样设备
dust-control techniques 粉尘控制技术

Exercises

4-1 大气污染物控制基于哪些原则？
4-2 扇风机遵循哪些定律？
4-3 粉尘控制的基本方法有哪些？
4-4 cfm, fpm, gpm 分别代表什么单位？
4-5 表 4-2 中的 MeV 代表什么单位？

5 Rock Mechanics

本章介绍矿山岩石力学问题，包括岩石和岩石力学的概念、岩石的分类、岩石力学性质及其试验、现场测试方法与仪器、有色金属高温采场传热模型及其解法、注浆长锚索预控顶有色金属高温采场 ANSYS 三维仿真分析及有色金属高温采场岩体锚固机理——膨胀抬升锚索树根筋混凝土拱理论（简称胀锚拱理论）。

5.1 Rock and Rock Mechanics

From an engineering standpoint rock is two materials: (1) intact rock—that is, pieces or possibly bodies of relatively uniform rock of a single petrological type that do not contain gross mechanical defects of geological origin, such as faults or joints; and (2) in-situ rock—that is, a rock mass of sufficient size to contain a representative sample of these gross mechanical defects. The rock mass may be made up of more than one rock type. The characteristics of intact rock relate to processes, such as drilling, crushing, grinding or other mechanical operations, in which a large percentage of the created rock surface is fresh, whereas the properties and behavior of in-situ rock relate to problems in designing or evaluating the stability of slopes or underground openings. The mechanical behavior of the two is so different that each type will be considered separately.

Rock mechanics is concerned with the mechanical properties and behavior of rock—that is, how rock behaves when subjected to a force field. The field may be induced by excavating a void in a rock continuum or produced by some mechanical means, such as drilling or crushing. This subject is of especial importance in mining because rock is both the principal construction material and the principal product of the excavation process.

5.2 Identification and Classification of Rock

5.2.1 Intact Rock

Geological identification of intact rock types usually is based on examination of short sections taken from drill cores, or, if underground access is possible, from selected pieces of rock. These identifications can be used to classify rock in the conventional way: that is, as igneous, metamorphic or sedimentary. Core examinations can be supplemented with borehole observations made with either an optical or television borehole camera. This class of information is especially helpful if sections of the rock mass are altered or decomposed to the degree that they do not core. In addition, mechanical property determinations should be made to measure the intact rock strengths, deformation moduli and frictional characteristics. These can be used to mechanically classify intact rock. For example, the system of classification by Deere and Miller (Table 5-1) utilizes the uniaxial compressive strength, C_0, and the tangent modulus of elasticity at 50% of C_0, $E_{t(50)}$. Rocks are classified as BM, AH, BL, etc.

Table 5-1 Engineering classification for intact rock①

	Class	Description	Uniaxial compressive strength, MPa
On basis of strength	A	Very high strength	>220
	B	High strength	110—220
	C	Medium strength	55—110
	D	Low strength	28—55
	E	Very low strength	<28
	Class	Description	Modulus ratio
On basis of modulus ratio	H	High modulus ratio	>500
	M	Average (medium) modulus ratio	200—500
	L	Low modulus ratio	<200

①After Deere and Miller. Modulus ratio = $E_{t(50)} / C_0$.

5. 2. 2 In-situ Rock

Mechanical evaluation and classification of in-situ rock usually is based on either an examination of exploratory drill cores or, if underground access is possible, from an inspection of the in-situ rock. Mechanical defects of in-situ rock are defined as follows:

(1) Fracture——A break in a body of rock which may or may not be attended by a displacement. Fractures may be of geological origin, or they may be man-made as, for example, those caused by blasting.

(2) Fault——A fracture of geological origin in a body of rock attended by a shear displacement. Faults usually contain gouge, decomposition or alteration products.

(3) Joint——A fracture of geological origin in a body of rock not attended by a displacement. Joints usually occur in sets or systems. There may be alteration or decomposition products on the joint surfaces.

(4) Parting——A thin depositional layer, often organic, separating thicker rock strata.

(5) Bedding Plane——The surface that separates each successive layer in a stratified body of rock.

In logging mechanical defects it is especially important to record the estimated degree of bond—that is, the estimated tensile strength—across each type of defect. Generally, the bond across a fault or fault zone is negligible, whereas the bond across a joint may vary from zero to a strength as great as, or greater than, that of the host rock. The bond across partings or bedding planes is variable, usually ranging from zero to a strength less than that of the host rock.

5.3 Mechanical Properties and Behavior of Rock

There are three characteristics of rock of especial concern to mining engineers: the deformational behavior—that is, how the rock deforms under the action of applied loads; the strengths—namely, compressional, tensile and shear; and the frictional properties—that is, how the rock yields on a frictional surface. In varying degree, these characteristics are time-dependent. However, tests performed in fractions of a second to minutes are referred to as time-independent or static, even though there is some time-dependent effect. Tests over longer periods (hours to years) are referred to as time-dependent and involve such phenomena as creep. Also, other physical or chemical effects, such as moisture or pore water pressure, may affect the properties of rock. In this subsection, the various properties of intact and in-situ rock are examined.

Static Properties (Intact Rock)—If a stress (force per unit area) is applied to any real material (for example, rock) it will deform. To achieve understanding of the deformational behavior of a real material, the deformational behavior of an idealized (hypothetical) material will be considered first. The stress-strain diagram in Fig. 5-1, is for a linear or perfectly elastic material: that is, a material for which $\varepsilon = \sigma / E$, where the constant of proportionality, E, is Young's modulus. No real material is perfectly elastic, although some igneous rocks approach this state. The mechanical model for a perfectly elastic material is a spring.

A perfectly plastic material will not deform if the stress is less than σ_0, it will deform permanently without limit if the stress is equal to σ_0, and it will not support a stress greater than σ_0 (Fig. 5-2). The mechanical model for a perfectly plastic material is a frictional contact. An elastoplastic material is linear-elastic to some stress level, σ_0, and perfectly plastic at that stress level

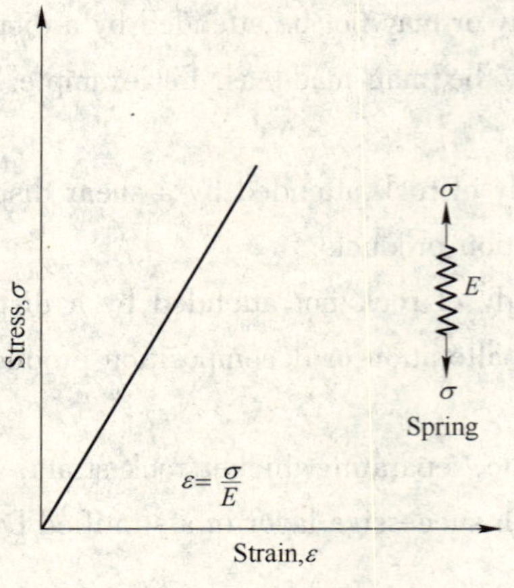

Fig. 5-1 Perfectly elastic material

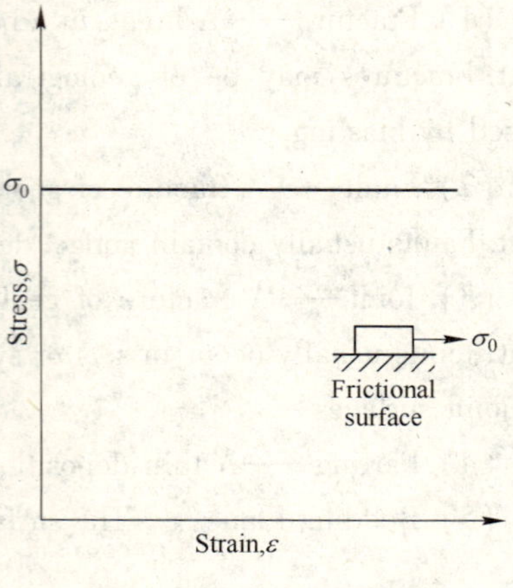

Fig. 5-2 Perfectly plastic material

(Fig. 5-3). The mechanical model for an elastoplastic material is a frictional contact in series with a spring.

The stress-strain curve in Fig. 5-4 represents a material that is linear-elastic to some stress level σ_1 (Point *a*). Above this stress level the material is inelastic: that is, the strain is not proportional to the stress. If, at some stress level σ_2 (Point *b*), the stress is removed and the strain vanishes, the material is termed elastic. Point b is referred to as the yield point, and σ_2 is the yield stress. If, at a stress level above σ_2, the stress is removed (say at Point *c*) and the strain does not vanish, the material is said to have developed a permanent set. If the stress is reapplied and the stress-strain curve is (approximately) linear from ε_1 to Point *d*, the material is said to have strain-hardened. If the stress then is increased above σ_3, the material will continue to strain inelastically to failure. Materials that deform past Point b are termed ductile.

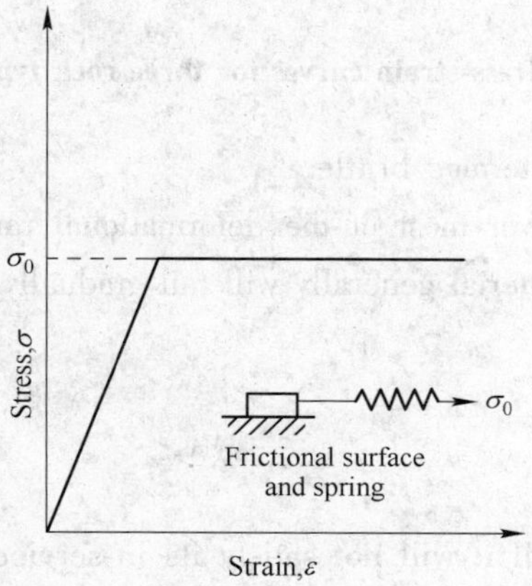

Fig. 5-3 Elastoplastic material

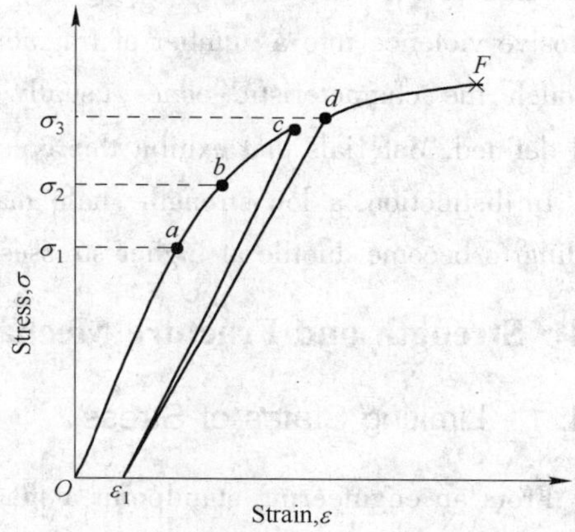

Fig. 5-4 Elastic-ductile material

The deformational behavior of an idealized rock material is illustrated in Fig. 5-5. The tangent modulus of elasticity, E_t, is the slope of the stress-strain curve at some specified stress: that is, $(\Delta\sigma/\Delta\varepsilon)_{\sigma_1}$. The secant modulus of elasticity, E_s, is the stress divided by the strain at some specified stress: that is, $(\sigma/\varepsilon)_{\sigma_2}$. Generally, the slope of the stress-strain curve for rock will increase at low stress levels, then become linear up to some stress level where it will decrease and, under certain conditions, become zero and then turn negative. The extent of the negative portion of the curve depends upon the "stiffness" of the testing machine, the rock type and the rate of loading. The negative slope indicates that some rock materials have ability to sustain a load even after the maximum stress level has been reached. It is generally presumed that at about the stress level

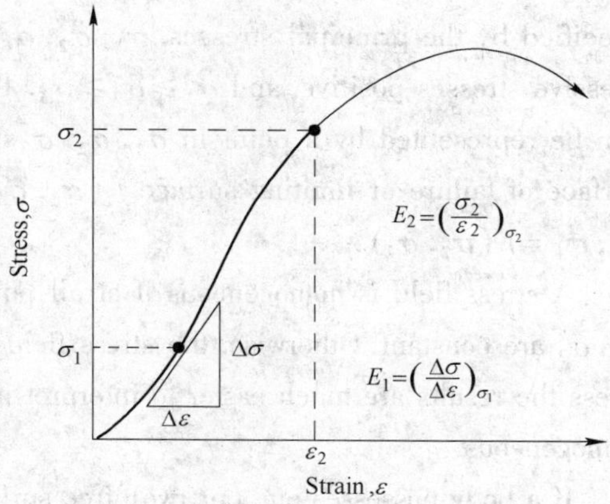

Fig. 5-5 Idealized stress-strain curve

where the slope of the stress-strain curve starts to decrease, the material is internally failing by the development or extension of micro- or macro-fractures.

The stress-strain curves for three rock types tested in uniaxial compression are given in Fig. 5-6. The limiting stress value attained, indicated by the points F, is referred to as the uniaxial compressive strength, C_0. Granite and other igneous rocks tend to show only a small deviation from linear elasticity up to failure. This class of rock usually breaks with almost explosive violence into a number of fragments, although the characteristic cones usually are well defined. Materials that exhibit this type of behavior are termed brittle.

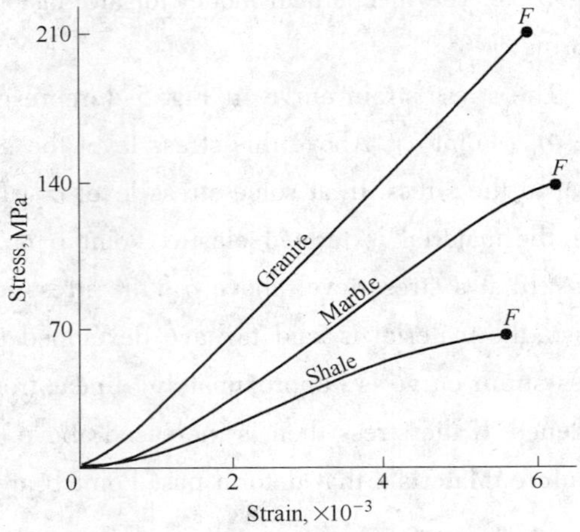

Fig. 5-6 Stress-strain curves for three rock types

In distinction, a low-strength shale may be inelastic over most of the deformational range, tending to become ductile at higher stresses. This type of material generally will fail gradually.

5.4 Strength and Fracture Mechanics

5.4.1 Limiting States of Stress

From an engineering standpoint a material has failed if it will not satisfy its in-service requirements. A brittle material may fail by the development or extension of cracks, or by fracturing into pieces where there is no cohesion across the fracture surfaces. A material also may fail because it deforms excessively, which, in a ductile material, may be accompanied by rupture surfaces—that is, surfaces on which there are displacements but not complete loss of cohesion. Whether by fracture or excessive deformation, mechanical failure results from some combination of applied stresses which produces in each element of the material a state of stress which can be specified by the principal stresses, σ_1, σ_2, σ_3. The convention employed in this section is compressive stresses positive, and $\sigma_1 \geqslant \sigma_2 \geqslant \sigma_3$. The set of σ_1, σ_2, σ_3 values for which failure occurs can be represented by a point in σ_1, σ_2, σ_3 space, and the totality of these points describes the surface of failure or limiting surface $f(\sigma_1, \sigma_2, \sigma_3) = 0$, or in terms of the dependent variable, σ_1, $\sigma_1 = f(\sigma_2, \sigma_2)$.

A stress field is homogeneous if at all points in a body the magnitude and directions of σ_1, σ_2, σ_3 are constant. Otherwise, the stress field is inhomogeneous. In determining limiting states of stress the results are much easier to interpret if the stress field in the specimen is (approximately) homogeneous.

If a body possesses one (or two) free surfaces—that is, surfaces across which one of the principal stresses is zero—it is referred to as unconfined. Most unconfined rocks fail as a brittle material by fracturing. In engineering applications rock generally is unconfined in at least one, and often two of the three, orthogonal directions, as, for example, in crushing, grinding or drilling rock, or on the surface of a mine pillar or other underground surfaces. Therefore, in service, most rocks behave as a brittle material and fracture is the usual mode of failure. However, there are exceptions, such as halite and other evaporite minerals, or rocks containing clay components. For these materials the limiting surface is indefinite.

In determining the limiting states of stress in rock, a strictly homogeneous stress field is difficult to achieve. The nearest approach is developed in tensile testing, especially if "dog bone" shaped specimens are employed. Nevertheless, it is customary to refer to tests in which external uniform loads are applied to solid prisms or cylinders of rock as homogeneous stress field tests, although, in some instances, deviations from homogeneity may be more than 20%.

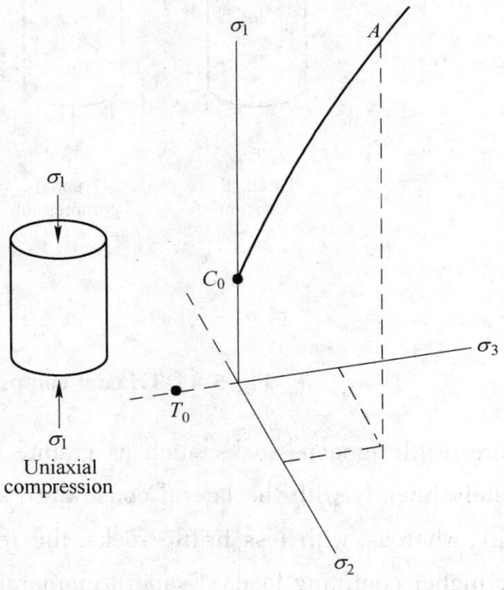

Fig. 5-7 Limiting states of stress

5.4.1.1 Uniaxial Tension($\sigma_1 = \sigma_2 = 0 > \sigma_3$)

The uniaxial tension test gives one point, T_0, on the failure surface $f(\sigma_1, \sigma_2, \sigma_3) = 0$ (Fig. 5-7). Relatively isotropic rock specimens subjected to uniaxial tension will fracture on a plane approximately normal to the axis of the specimen, producing extension or cleavage fracture surfaces.

5.4.1.2 Uniaxial Compression($\sigma_1 > \sigma_2 = \sigma_3 = 0$)

The uniaxial compression test gives one point, C_0, on the limiting surface, $f(\sigma_1, \sigma_2, \sigma_3) = 0$ (Fig. 5-8). In most uniaxial compression tests the specimen fails on conical (shear) surfaces that exhibit considerable powdering and abrasion of individual crystals. With igneous rocks, failure usually is sudden and explosive, whereas with many lower strength sedimentary rocks, and with evaporite minerals, failure may be less sudden, with some indication of yield prior to failure.

The mode of failure is affected by the type of loading machine employed. With most conventional hydraulic or screw-type machines, when initial yield and/or failure starts the energy stored in the machine is transmitted to the specimen, accelerating the failure processes. However, in "rigid-type" compression machines, in which the energy stored in the machine is small and, hence, the energy transmitted to the specimen is greatly reduced, it has been found that some specimens yield to the degree that the slope of the stress-strain curve may become negative prior to failure.

5.4.1.3 Triaxial Compression($\sigma_1 > \sigma_2 = \sigma_3 > 0$)

The limiting state of stress for the triaxial test is represented by a line, C_0A (Fig. 5-7). For the

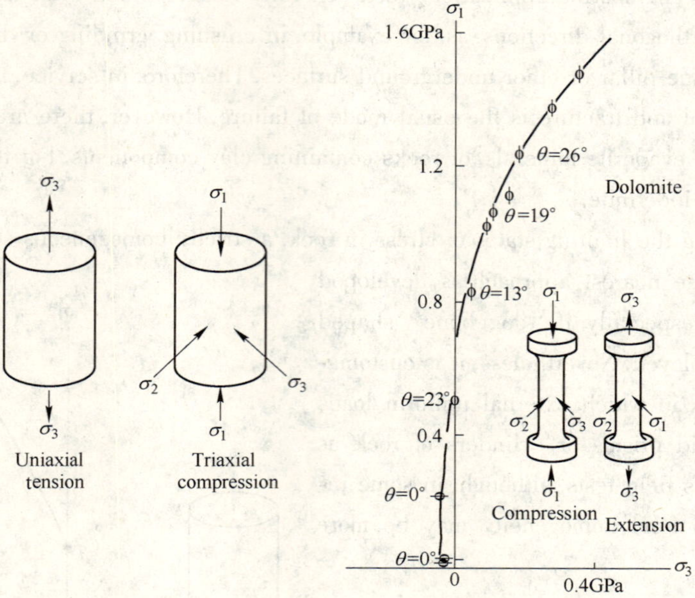

Fig. 5-8 Triaxial compression-extension results for dolomite

more brittle igneous rocks, such as granite or diabase, the limiting axial stress increases approximately linearly with the lateral constraint (at least in the load range normally experienced in mining), whereas, with less brittle rocks, the relationship generally deviates from linearity, especially for higher confining loads. Evaporite minerals and carbonates may not fail by fracture but rather by flow or rupture, sometimes under relatively low constraint. The rupture surfaces occur on a diagonal plane whose normal makes an angle less than 45° with the axis of the specimen.

5.4.1.4 Triaxial Extension($\sigma_1 = \sigma_2 > \sigma_3$)

This test is effected with "dog bone" shaped specimens but otherwise triaxial conditions apply. Because the ends of the specimens have a larger area than the center sections, the hydraulic load produces both a radial compression and an axial tension in the specimens. Brace found the σ_1, σ_3 curve for the transition from triaxial compression to extension to be continuous (Fig. 5-8). In compression, the fracture surfaces were inclined to the specimen axis and exhibited shear characteristics. In tension, cleavage fracture surfaces normal to the specimen axis were created.

5.4.1.5 Tests in Inhomogeneous Stress Fields

Many tests have been performed on specimens in which the stress field was inhomogeneous, using beams, plates, hollow cylinders, etc. In some of these tests the object was to determine the effects of the intermediate principal stress, σ_2, which the triaxial test does not permit.

In general, it has been found that the intermediate stress affects the limiting stresses, σ_1 and σ_3, by less than 20%. From a practical standpoint, because neither the in-situ state of stress nor the rock strength can be measured within this limit, the effect of the intermediate stress can be neglected.

5.4.2 Mechanics of Brittle Fracture

The mechanics of brittle fracture is concerned with what takes place within a material prior to and during failure, as opposed to strength tests, which only determine limiting states of stress. The strength of solids has been estimated on the basis of the bonds between atoms and molecules, with the result that mechanical strengths two or three orders of magnitude greater than actual values were calculated. To obtain more realistic strength values Griffith postulated that unstressed material (glass) contains microscopic cracks (flaws) around which stress concentrations develop when a material is stressed. On the basis of theory developed by Inglis, the stress concentration around an elliptical crack in an isotropic material was determined. In a tensile stress field a crack whose axis is normal to the applied stress (Fig. 5-9) will develop a tensile stress concentration at the crack tip whose magnitude is proportional to the crack length and inversely proportional to the radius of curvature of the crack tip. As the radius of curvature approaches an atomic diameter the magnitude of the crack tip, tensile stress can approach the strength of atomic bonds. Hence the crack should extend and ultimately cause a fracture surface normal to the applied stress to develop. This tensile failure mechanism is consistent with observation.

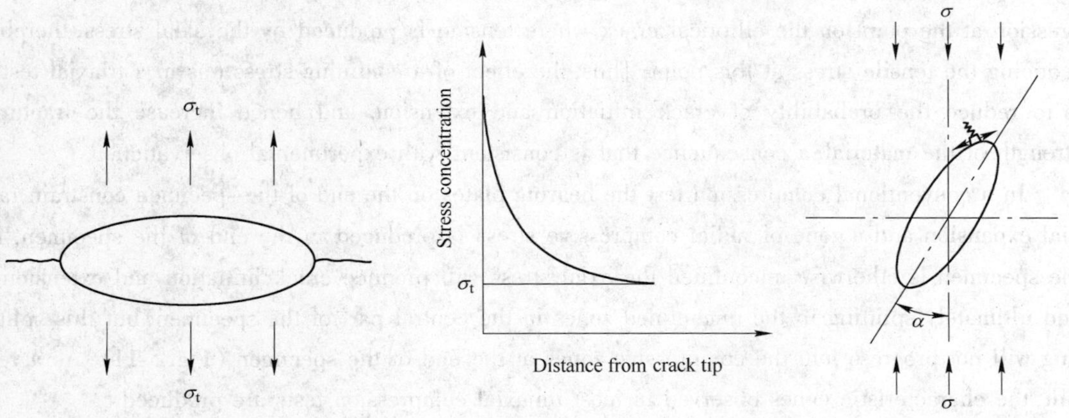

Fig. 5-9 Elliptical crack in a tensile stress field

Griffith further considered the stress concentration around an elliptical crack in a compressive stress field. If the crack is inclined at an angle, α, with respect to the applied stress, a tensile stress will develop on the crack surface as indicated in Fig. 5-9. This tensile stress has a maximum near the crack tip and, when it exceeds the tensile strength of the rock, a crack will develop. Thus, a tensile crack can develop and extend in a compressive stress field. This branch crack can in turn develop a branch crack, and so on, thus extending fracture in the direction of the applied compressive stress (Fig. 5-10). The same mechanism can be shown to develop around an infinitely thin crack.

The Griffith fracture mechanism assumed that the elliptical crack is in an isotropic elastic material and that the stress field is homogeneous. On the scale of crystal or grain size, rock is neither isotropic elastic, nor is the stress field homogeneous. Therefore, it is unlikely that an initial branch crack would continue to branch until a fracture surface develops. To account for the development of a fracture surface it is necessary to assume that rock contains a large number of microcracks, and that these cracks can have different lengths and orientations. Then, as a specimen is loaded, the microcrack with the most unfavorable length and orientation will produce a branch crack, which may or may not produce additional cracking but will not lead to the development of a fracture surface. Then, the microcrack with the next most unfavorable length and orientation will develop a branch crack, etc. Ultimately, these extended cracks will join or coalesce with neighbors until a fracture surface forms. This procedure should lead to the development of a fracture surface lying in the plane of the applied (uniaxial) compressive stress. This type of "splitting" is observed in compressive test specimens in which the end constraint on the specimen has been minimized with friction reducers so that the applied stress field is essentially uniaxial and homogeneous.

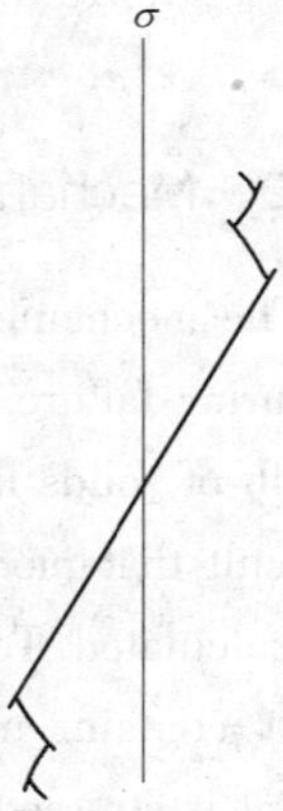

Fig. 5-10 Branching crack

If, in addition to the axial stress, a lateral stress is applied to a specimen it will produce compression at the point on the elliptical crack where tension is produced by the axial stress, thereby reducing the tensile stress at this point. Thus, the effect of a confining stress (as in a triaxial test) is to reduce the probability of crack initiation and extension, and hence increase the fracture strength of the material, a consequence that is consistent with experimental observation.

In a conventional compression test the bearing plates on the end of the specimen constrain radial expansion and a zone of radial compressive stress is produced in the end of the specimen. If the specimen is otherwise unconfined the axial stress will produce crack initiation and extension, and ultimately splitting in the unconfined zones in the central part of the specimen, but this splitting will not progress into the compressive zones at the end of the specimen (Fig. 5-11). As a result, the characteristic cones observed in most uniaxial compression tests are produced.

Because microcracks have both a size and an orientation distribution, the mechanism of fracture is a statistical process. Also, because the larger the specimen the greater the probability it will contain a "worst" crack, the strength required to produce fracture should depend on the specimen size. The "size effect" has been the subject of numerous investigations. For cubical specimens of coal with side length, L, the relationship $C_0 = kL^{-1/2}$, where C_0 is the uniaxial compressive strength, has been found valid within the range of 1, 170 to 3, 450Pa and $L \geqslant 1,625$mm.

Stress gradients result form stress concentrations produced by either nonuniform boundary loading as, for example, a center-loaded beam, or from boundary irregularities, such as a notch or an internal hole in an edge-loaded plate. Stress gradients increase as loaded boundary areas or the radii of curvature of notches, internal holes, etc., decrease. Thus, in a test specimen, the volume of

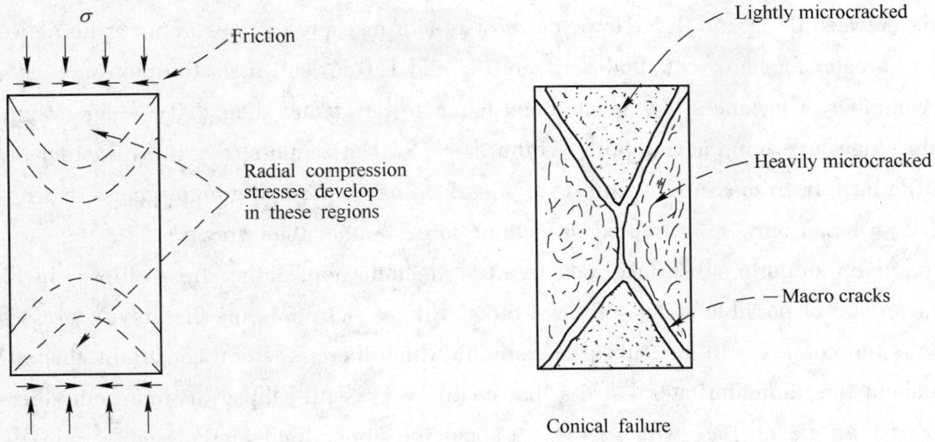

Fig. 5-11 Compression specimens with unlubricated ends

a material in which a critical (failure) stress develops depends on how it is loaded and its internal or external boundary shape, or both. That is, this stress gradient effect can be considered as a specimen size effect.

The stress gradient effect is exemplified in the flexural test, in which a high tensile stress gradient is produced on the opposite side of a center-loaded beam directly under the load point. The flexural test gives a calculated tensile strength two to three times the uniaxial tensile strength. Also, Durelli et al. reported the calculated shear strength in the kerf area produced by model diamond drill to be about one order of magnitude higher than the triaxial shear strength for the same material.

5.5 Stress in Rock

If design or evaluation of the stability of a surface or subsurface opening or system of openings is to be made on a rational basis, knowledge of the stress field—that is, the state of stress in rock preexisting excavation, and the stress concentrations that develop in the rock surrounding an opening—is required. Some inferences regarding the subsurface state of stress can be made by considering the gravitational force acting on the incumbent rock. Techniques are available for determining the stress concentration and stress field from measurements made in the walls of subsurface openings. Also, some inferences can be made from hydraulic fracturing tests in deep wells and form core discing when it occurs. From these sources, generalizations regarding the subsurface state of stress can be made.

From a consideration of the gravitational forces acting on an element of volume in an elastic sphere (for example, the Earth) Phillips has shown that the state of stress at depth can be approximated by $S_v = \gamma h$ and $S_h = \sigma_v \left(\frac{\nu}{1-\nu}\right)$, where S_v and S_h are the vertical and horizontal components of stress, h is the vertical depth, ν is Poisson's ratio, and γ is the unit weight of the rock. Poisson's ratio for most rocks is between 0. 2 and 0. 33; hence, the ratio of the horizontal to vertical stress should lie between 0. 25 and 0. 5. However, most in-situ measured values of S_h/S_v lie between 0. 5 and 1. 5 or greater for hard rock, and between 0. 8 and 1. 0 for soft inelastic rock, such as shale or salt. In a number of instances, S_h/S_v has been found to be greater than 1. 0——for example, in a flat-bedded limestone mine at a depth of 700m $S_h = 2S_v$. Horizontal stresses of the order of several to tens MPa have been measured within 15m of surface in granite and granite-gneiss quarries. A ratio of $S_h/S_v > 1$ can only be attributed to tectonic forces (other than gravity).

In relatively uniform subsurface rock free from tectonic forces, the stress ellipses in Fig. 5-12 are characteristic of possible states of stress. Stress Ellipse Ⅰ represents the gravity stress field for a Poisson' s ratio of 0. 25. In any subsurface area in which there is lateral constraint, this ellipse represents about the minimum lateral stress that could be present. Ellipse Ⅱ represents the state of stress for (1) an elastic rock with ν = 0. 4 (about the upper limit for Poisson' s ratio); (2) an elastic rock in which plastic or viscoelastic deformations have permitted some equalization of the vertical and horizontal stresses; or (3) an elastic or inelastic rock with ν < 0. 4, acted on by a horizontal tectonic force. Ellipse Ⅲ (circle) represents a state of stress in an inelastic rock, such as salt at great depth in which plastic or viscoelastic effects have permitted a complete equalization of the vertical and horizontal stresses, or in any elastic or inelastic rock acted on by horizontal forces of a magnitude such that $S_v = S_h$. Ellipse Ⅳ depicts the state of stress in either an elastic or inelastic rock acted on by a horizontal compressional force. The distinction between Ellipses Ⅱ and Ⅲ and Ellipse Ⅳ is that the horizontal compressional force is necessary in the latter case.

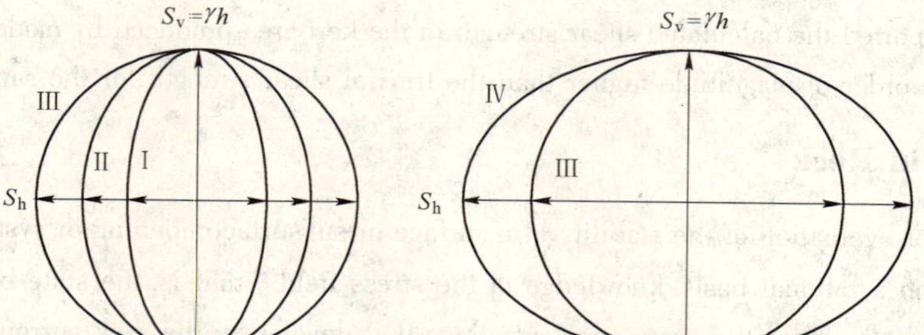

Fig. 5-12 Stress ellipses

The subsurface state of stress may be affected by the presence of large-scale mechanical anomalies. For example, consider a hypothetical body of rock (hatched area Fig. 5-13) which crystallized after the surrounding area had solidified, and in the process of crystallization expanded uniformly in all directions, placing the body of rock in hydrostatic compression. If, in addition, this body of rock is in a gravity stress field, both the horizontal and vertical components of stress at Point A would be greater than those for a gravity stress field alone, as illustrated in Ellipse Ⅱ, Fig. 5-13, where Ellipse Ⅰ represents the gravity stress field. At Point *B* in the stress concentration zone outside the compressed body of rock, the vertical stress would be less than, and the horizontal stress greater than, the corresponding gravity components, as indicated by Ellipse Ⅲ. Also, because

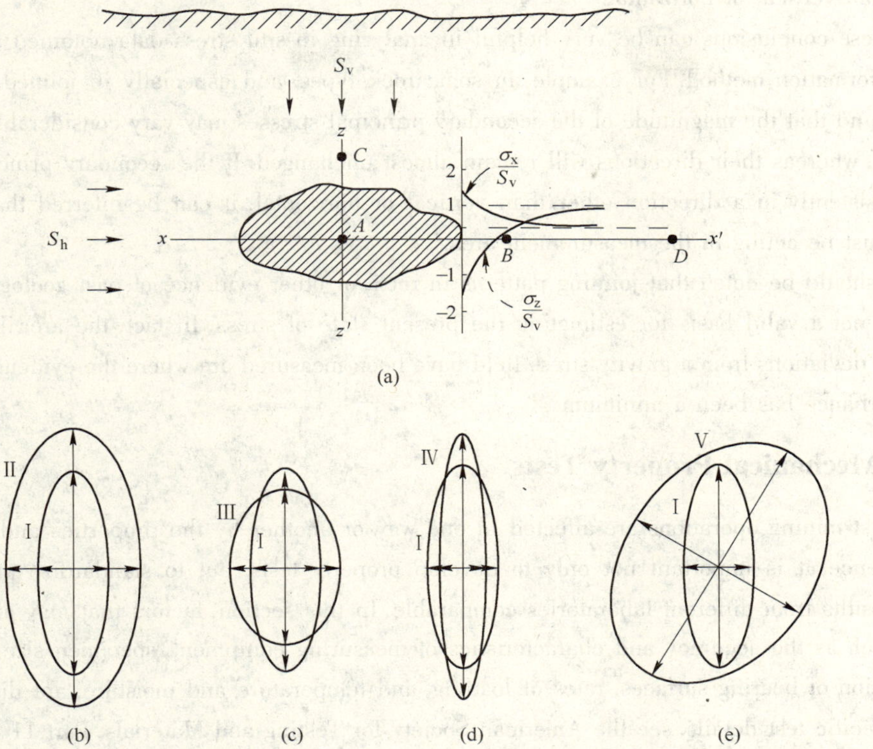

Fig. 5-13 Inclusion in a gravity stress field

of equilibrium requirements, the vertical stress averaged along any horizontal line through the area and adjoining stress concentration zones (as along the line x, x') would be $S_v = \gamma h$. Correspondingly, at Point C, the horizontal stress would be less than, and the vertical stress greater than, the corresponding gravity components, as indicated by Ellipse Ⅳ. If the depth of the anomaly from surface is much greater than its vertical dimension, the horizontal stress averaged along any vertical line through the anomaly and its stress concentration zones would be S_h.

If the stress within the anomaly is not uniform, but such that the larger secondary stress lies in some direction other than the horizontal or vertical, the stress ellipse could be like that in Ellipse Ⅴ. However, the vertical stress averaged along any horizontal line through the area would remain $S_v = \gamma h$, and the horizontal stress averaged along any vertical line through the are would remain S_h.

Although this example is hypothetical it illustrates two important points. First, because the surface is free to move (is unconstrained), the vertical stress, averaged over any equidepth (h = constant) plane of sufficient area, will be γh. The area over which the average must be taken will depend on the size of the anomaly. Geological information is insufficient for estimating this area, but for larger geological features it might be of the order of square km. Second, tectonic forces acting in the horizontal direction will affect the horizontal stress, but the vertical stress will remain unaffected. Tectonic forces acting in any direction other than vertical or horizontal will affect both the vertical and horizontal components of stress, and the axes of the stress ellipse will be in some direction other than vertical or horizontal.

These conclusions can be very helpful in analyzing in-situ stress data obtained by the borehole-deformation method. For example, in some rock types, and especially in jointed rock, it has been found that the magnitude of the secondary principal stresses may vary considerably form point to point, whereas their directions will remain almost unchanged. If the secondary principal stresses are consistently in a direction other than vertical or horizontal, it can be inferred that a tectonic force must be acting in the measurement area.

It should be noted that jointing patterns in rock, or other evidence of past geological disturbance, is not a valid basis for estimating the present state of stress. In fact, the area in which the greatest deviations from a gravity stress field have been measured are where the evidence of geologic disturbance has been a minimum.

5.6 Mechanical Property Tests

Most mining operations are affected in one way or another by the properties and behavior of rock. Hence, it is important not only to develop property tests, but to standardize procedures to make results from different laboratories comparable. In this section, factors that may affect test results, such as the accuracy and characteristics of measuring equipment, specimen size and shape, preparation of bearing surfaces, rates of loading, and temperature and moisture, are discussed. For more specific test details, see the American Society for Testing and Materials, Part 11 (which prescribes standardized tests for soil and rock).

5.6.1 Definitions and Concepts

A number of terms with specific meanings are sued in the description and specification of instruments and procedures for measuring stress, strain and displacement (or deformation). These are defined and explained in the following paragraphs.

(1) Deformation (or Displacement) Sensitivity——Deformation sensitivity is defined as the smallest increment of the measured quantity that can be read on the indicating scale. Many instruments employ magnification—mechanical, optical, electronic amplification, etc.—to improve the sensitivity. If the smallest increment of length that can be read on the scale of a deformation device is d, and the magnification is m, the deformation sensitivity of the device, S_d, is $S_d = d/m$. As a general rule in selecting an instrument, the deformation sensitivity should be about 1% of the anticipated range of measurement.

(2) Strain Sensitivity——If a displacement-measuring device is used as a strain gage—that is, if the displacement is divided by the gage length, L (sometimes referred to as the base length)—the strain sensitivity of the instrument, S_s, is defined as $S_s = d/(mL)$.

(3) Range——In selecting a deformation or strain gage, the range of the instrument must be considered. It is defined as the maximum quantity the instrument can measure. Usually, if the range of an instrument is increased, the deformation sensitivity is correspondingly decreased.

(4) Drift or Time Stability——If a measurement is made over an extended period the meas-

uring instrument should not drift during this period. Drift, or time stability, refers to the change in the reading of an instrument not caused by a change in the effect being measured or by factors normally controlled during the course of measurement.

(5) Reproducibility and Accuracy——In a succession of identical observations, an instrument will not indicate exactly the same reading—that is, the readings are not exactly reproducible. The most probable value of an experimentally determined quantity is the arithmetic mean. Note that neither reproducibility nor sensitivity is any indication of the accuracy of the instrument, which can be determined only by comparing it against some know standard.

5.6.2 Laboratory Instrumentation

Because the breaking strain for rock is small (about 0.001 to 0.003 in tension and 0.002 to 0.010 in compression) instruments for measuring deformation or strain should be very sensitive. A number meeting this requirement are commercially available and, with only minor modification, can be used for rock-mechanics tests.

(1) Deformation Gages——The characteristics of three types of deformation gage are given in Table 5-2. Micrometers (micrometer heads) are simple mechanical devices with sufficient sensitivity for many rock mechanics tests. They have a large range, are rugged, are relatively temperature-independent, and are free from drift. Dial indicators are more sensitive than micrometers, relatively temperature independent, and free from drift. Linear variable differential transformers (LVDTs) convert linear displacement into an electrical analogue. They come in a variety of ranges, and, with proper circuitry, high sensitivities can be developed. LVDTs, in contrast to mechanical gages, permit remote sensing.

Table 5-2 Deformation gages

Gage type	Range, mm	Magnification	Deformation sensitivity, $\mu\varepsilon$	Temperature stability	Time stability
Micrometer (head)	25.40	60	100—1000	Excellent	Excellent
Dial indicator	2.54—25.40	100—1000	50—1000	Excellent	Excellent
LVDT	1.27—76.20	Variable	10—100	Fair	Fair

(2) Strain Gages——Strain gages differ from deformation gages in that they have a fixed gage length. Thus, direct reading or bridge indicator types can be calibrated to read directly in strain units (that is, in units of deformation per unit gage length). A number of the more common strain gage types are listed in Table 5-3, together with their characteristics. The Tuckerman optical strain gage consists of a collimating telescope and an optical strain gage. Bonded resistance strain gages are used effectively because they have both a high sensitivity and a comparatively large range (up to 2% strain). Also, they permit remote sensing. The circular photoelastic strain gage permits a determination of both the magnitude and direction of the stress in the plane of the gage.

(3) Loading Equipment——Universal testing machines, although expensive, provide for testing in both tension and compression. Load capacities are available from 27 to 2, 268t. Relatively inexpensive compression machines, designed for testing concrete cylinders, are available in load

ranges from 1.7 to 7.3GPa. For some tests, hydraulic jacks can be used for loading, and are available in load ranges from 9 to 90t. They are relatively inexpensive.

Table 5-3 Characteristics of typical strain gages

Gage type	Gage length, mm	Magnification	Range, mm	Strain sensitivity, $\mu\varepsilon$	Temperature stability	Time stability
Mechanical (whitmore)	50.8—254	1,000	2.54	10—50	Excellent	…
Optical (tuckerman)	12.7—101.6	2,500	0.1143	4	Excellent	…
Resistance① (unbonded)	25.4	2,000	0.0762	5	Good	Excellent
Resistance① (bonded)	3.2—50.8	2,500—5,000	0.0508—0.508	2—4	Good	Fair
Vibrating① wire (maihak)	101.6	10,000	0.1016	1	Good	Good
Photoelastic (linear)	25.4—50.8	…	…	70	Fair	Poor
Photoelastic (circular)	1.25*D*	…	…	50	Fair	Poor

①With companion bridge indicator.

5.7 Instrumentation and Procedures for Field Tests

Achieving any rational basis for designing an underground opening or system of openings, an open-pit or road cut, or a foundation for a surface structure, requires a knowledge of both the mechanical properties and the state of stress in the surrounding rock. This information, together with knowledge of the deformations produced by or following excavation, are essential for evaluating the stability of both surface and underground openings in rock. The development of instrumentation and procedures for measuring rock properties, the state of stress, and structural deformations has been the subject of many investigations. Some of the instruments and more generally accepted procedures are described in the following.

5.7.1 Instruments for Visual or Photographic Observations in Boreholes

In brief, these instruments are as follows:

(1) Stratoscope——The stratoscope is an optical periscope for visually or photographically examining the interior of a borehole. Stratoscopes are commercially available in diameters ranging form 38mm (to fit in an AX diamond drill hole) to 70mm (NX hole), and with the extendable lengths from 3 to 27m. Stratoscopes with lengths of 3 to 6m are particularly useful in observing the development of fractures or, when equipped with a reticle, in measuring the displacement across partings, joints, etc. From repeated measurements, the change in these displacements can be deter-

mined. Longer length stratoscopes are useful in making examinations in exploratory boreholes, especially if the recovered core is inadequate for this purpose.

(2) Borehole Cameras——Borehole cameras are used for photographing the interior walls of NX or larger diameter exploration holes. Cameras have been developed that can be lowered on a cable into vertical or near-vertical holes up to 305m or more deep. The instruments include a compass, which also is photographed so that the azimuthal orientation of the hole can be established. In a vertical hole, both the dip and strike of subsurface joints, bedding, etc. , can be established. Opportunity to employ color photography is an aid in identifying rock types.

(3) Borehole Television Cameras——Borehole television cameras are commercially available for lowering on a cable into vertical or near-vertical holes up to 305m deep. The image is transmitted through the lowering cable to the surface where it can be both visually observed and photographed. The cameras include a compass and an inclinometer for measuring both azimuthal orientation and inclination of the borehole. Television borehole cameras reproduce only in black and white.

5.7.2 Instruments for Measuring Deformation

The deformeters described in the following are employed for measuring deformations produced by excavation: for example, surface subsidence, roof convergence, deformations in the walls or structural parts of a mine, or deformation produced by artificial loading, as in a jacking test. Included are optical levels and theodolites, as they can be used for deformation measurements over large distances: for example, measurement of surface subsidence. A precision level used under favorable conditions can measure a vertical displacement of about 0.003m at 30m. By triangulation, horizontal displacements can be measured with about the same accuracy with a theodolite.

Rigid-type deformeters in lengths up to 3m or more are constructed from solid bars or tubes, with a micrometer head, dial gage or other type of sensing device attached to one end (Fig. 5-14). This type of instrument generally is used to measure surface deformations of the walls or other components of underground structures. If a rigid deformeter is to be used where there is a large temperature variation, the rod or tube should be of invar steel. Measurements between fixed pins mounted on the surface of an underground opening normally can be reproduced to within ±0.05mm.

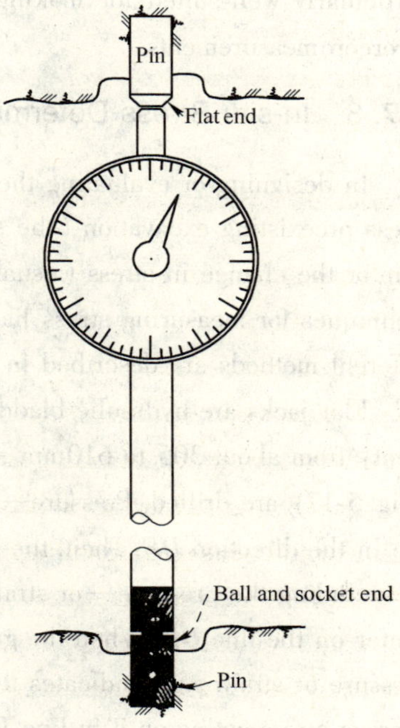

Fig. 5-14 Rigid-type deformeter

Rigid-type deformeters in lengths up to 9m or more also are used to measure axial deformation in boreholes. For this measurement one end of the deformeter is anchored in the end of the borehole (Fig. 5-15). The deformation is made between the other end of the deformeter and a reference point on the collar of the hole. Because this type of instrument is

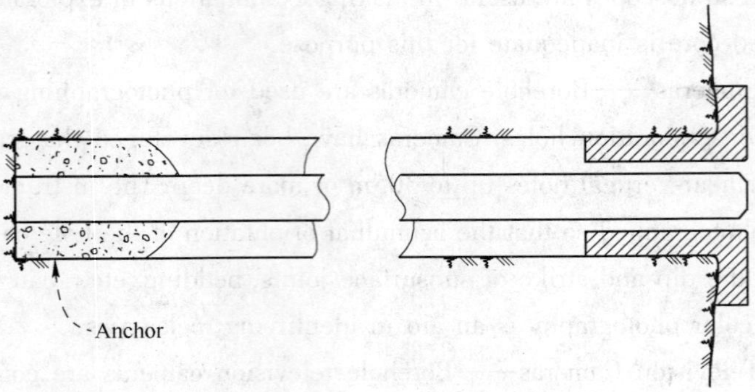

Fig. 5-15 Borehole deformeter

left in position, and because the temperature variation in a borehole in rock is small, accuracies of ±0.025mm normally can be obtained.

Axial-type multiple-position borehole deformeters are used for measuring displacements over the length and between points within a borehole as, for example, across an open joint. A typical instrument consists of seven rods anchored at selected points within a borehole. The rods extend to the collar of the hole, and deformation measurements are made between the ends of the rods and a reference point at the hole collar. To provide for remote sensing, linear variable differential transformers are used to measure the deformations.

Because axial borehole deformeters usually are left in place, and because of the small temperature variation in boreholes, the time stability of this class of instrument is good. Hence, they are particularly well-suited for making change-in-deformation measurements over long periods—that is, creep measurements.

5.7.3 In-situ Stress Determination

In designing or evaluating the stability of an excavation in rock, a knowledge of the absolute stress preexisting excavation (the stress field), the absolute stress distribution around the excavation, or the change in stress (usually with time) may be required. The development of gages and techniques for measuring stress has been the subject of extensive investigations. Several basically different methods are described in the following paragraphs.

Flat jacks are hydraulic bladders, usually made of malleable steel in sizes (for stress measurement) from about 305 to 610mm square (Fig. 5-16). To measure absolute stress, Holes *A* and *B* (Fig. 5-17) are drilled. Pressure—or strain—indicating devices are inserted and oriented to measure in the direction *DE*. Then, the slot *CC′* is line-drilled, producing a partial stress relief in the area including the pressure—or strain—sensing devices. Next, the jack is grouted in the slot with its center on the line *DE*. When the grout has hardened the jack is pressurized to a value such that the pressure or strain gage indicates its initial value. The jack pressure is assumed to be equal to the average stress acting on it in line *DE*. Because time stability is excellent flat jacks also can be used to measure a stress change over extended periods.

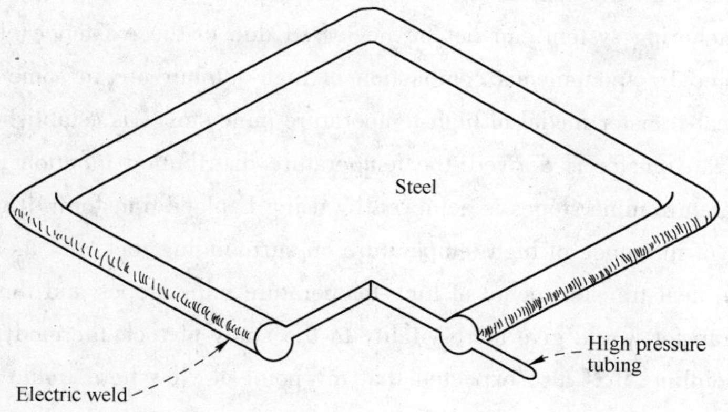

Fig. 5-16 Flat jack

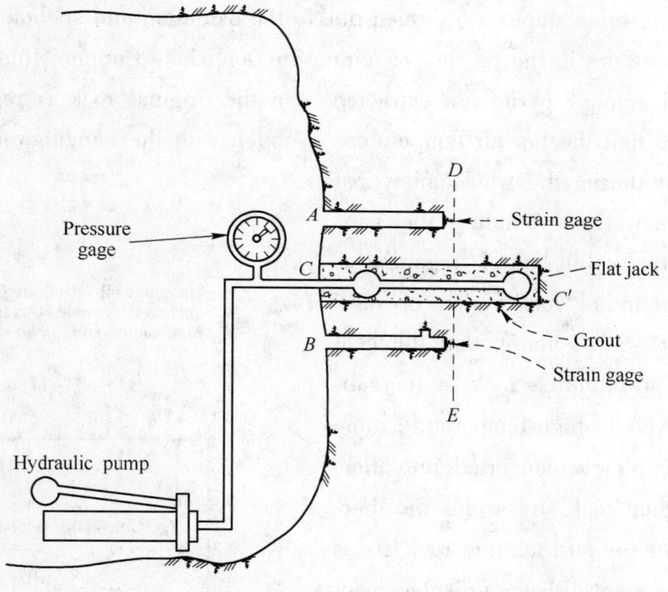

Fig. 5-17 Flat jack procedure

Although the flat jack technique is simple and the results easy to interpret, it has two serious disadvantages: (1) only stress acting in the direction normal to the flat jack can be determined; (2) because of installation difficulties only near-surface measurements can be made. Often the near-surface rock has been disturbed by blasting or stress readjustments.

5.8 Heat-transfer Model of High-temperature Non-ferrous Metal Mine Stopes and Its Solution

5.8.1 Introduction

Rock thermodynamics is a new borderline discipline. The theoretical study of it has just commenced, but it will find a wide application in the future. Influence of high temperature on surrounding rock and its anchoring system can not be neglected due to the existence of high-temperature mine stopes produced by spontaneous combustion of high-sulphur ore in some non-ferrous metal mines. Here, the heat-transfer model of high-temperature mine stopes is established; the differential equation of rock heat-transfer is derived; the temperature distribution function in the surrounding rock of high-temperature mine stopes is achieved by using Laplace transform. It is helpful and beneficial to the study of influence of high temperature on surrounding rock and its anchoring system. It is hoped that the heat-transfer model of high-temperature mine stopes and the differential equation of rock heat-transfer would give more vitality to the study of rock thermodynamics which is a new borderline discipline. It is also expected that my point of view here would have the effect of "throwing a brick to attract a piece of jade".

5.8.2 Heat-transfer Model of High-temperature Non-ferrous Metal Mine Stopes

High-temperature mine stopes are formed due to the oxidation and spontaneous combustion of colloidal sulphur pyrite ore in the process of mining in Dongxiang Copper Mine, Jiangxi Province, China. The colloidal sulphur pyrite ore extracted from the original rock is regarded as heat resource. It is assumed that the hot air temperature in touch with the hangingwall surrounding rock would keep constant during the whole analytical process. The hangingwall surrounding rock anchored by fully grouted cable bolts is regarded as the hangingwall surrounding rock system of high-temperature mine stopes, assuming that the heat in the hangingwall surrounding rock system absorbed from the hot air of high-temperature mine stopes would be completely converted into thermal stress. The original rock stress plus the thermal stress consists of the surrounding rock stress which makes the hangingwall have a displacement or even collapse and do mechanical expansion work W. Heat transfer is the main heat transmission type between the hangingwall surrounding rock and hot air in high-temperature mine stopes. Heat convection and radiation are secondary and can be ignored. The heat-transfer model of high-temperature mine stopes is illustrated in Fig. 5-18 in which t_w stands for rock surface temperature and t_f gives hot air temperature.

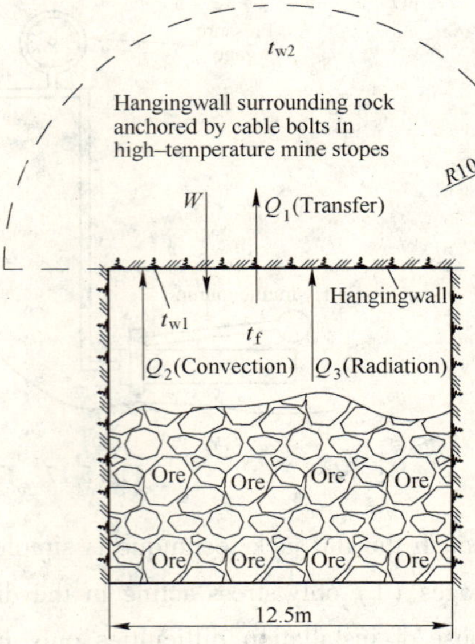

Fig. 5-18 Heat-transfer model of high-temperature mine stopes

5.8.3 Differential Equation of Rock Heat-transfer

When the temperature distribution in rock is uneven, an isothermal surface is composed of the points of the same temperature in rock. A bunch of isothermal lines is formed if isothermal surfaces intersect a plane. Thermal current flows along the direction perpendicular to isothermal lines (see Fig. 5-19).

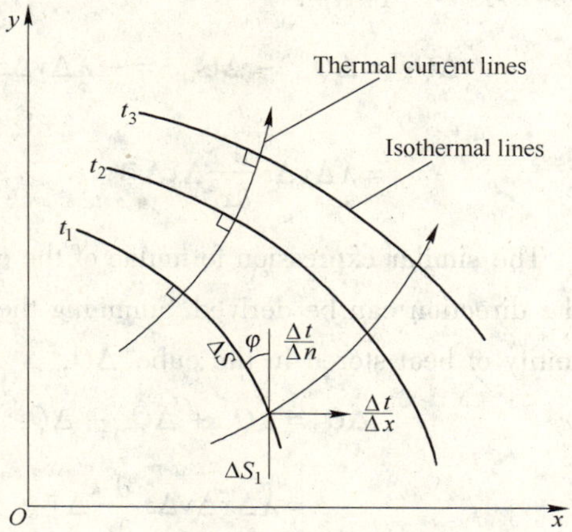

Fig. 5-19 Isothermal lines and thermal current lines

Let's analyze the little area ΔS in an isothermal surface. The quantity of heat ΔQ through ΔS in the time interval $\Delta\tau$ is expressed by the following equation

$$\Delta Q = -\lambda \Delta S \frac{\Delta t}{\Delta n} \Delta\tau \qquad (5\text{-}1)$$

where, $\frac{\Delta t}{\Delta n}$ is the temperature gradient along normal to isothermal lines and λ means heat-transfer coefficient, W/(m · K) or W/(m · ℃).

Suppose that $\frac{\Delta t}{\Delta x}$ is the temperature gradient along x direction. Let's consider the quantity of heat ΔQ_x through ΔS_1 in the time interval $\Delta\tau$. The surface ΔS_1, which intersects the isothermal line to form an angle φ, is perpendicular to x direction. Due to that $\Delta S_1 = \Delta S/\cos\varphi$ and $\frac{\Delta t}{\Delta x} = \frac{\Delta t}{\Delta n}\cos\varphi$, we have the equation

$$\Delta Q_x = -\lambda \Delta S_1 \frac{\Delta t}{\Delta x} \Delta\tau \qquad (5\text{-}2)$$

Assume the rock body in Fig. 5-20 to be homogeneous. In order to derive the differential equation of rock heat-transfer, the little cube $\Delta x\Delta y\Delta z$ which is in Cartesian coordinate system is took from the rock body. If the temperature at the cube center is t, the temperature at all points on left side surface of the cube can be considered identical and equals $t - \frac{\partial t}{\partial x} \cdot \frac{\Delta x}{2}$ in which $\frac{\partial t}{\partial x}$ is the temperature gradient along x direction. We can also get the temperature t_{xR} on right side surface of the cube, $t + \frac{\partial t}{\partial x} \cdot \frac{\Delta x}{2}$ · From the equation (5-2), We get that $\Delta Q_{xL} = -\lambda \Delta y \Delta z \frac{\partial}{\partial x}\left(t - \frac{\partial t}{\partial x} \cdot \frac{\Delta x}{2}\right)\Delta\tau$ and $\Delta Q_{xR} = -\lambda \Delta y \Delta z \frac{\partial}{\partial x}\left(t + \frac{\partial t}{\partial x} \cdot \frac{\Delta x}{2}\right)\Delta\tau$ where ΔQ_{xL} and ΔQ_{xR} are the quantity of heat flowing into the left side surface and that flowing out of the right side surface, in the time interval $\Delta\tau$ re-

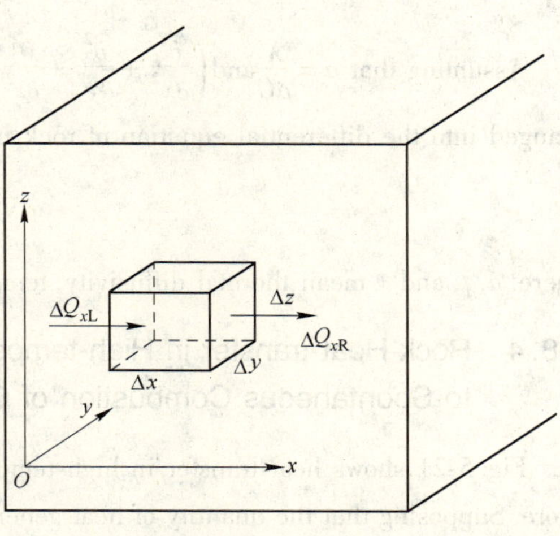

Fig. 5-20 Heat-transfer of the little cube

spectively. The quantity of heat consumed in the cube along x direction is showed as follows:

$$\Delta Q_x = \Delta Q_{xL} - \Delta Q_{xR} = -\lambda \Delta y \Delta z \left\{ \left(\frac{\partial t}{\partial x} - \frac{\partial^2 t}{\partial x^2} \cdot \frac{\Delta x}{2} \right) - \left(\frac{\partial t}{\partial x} + \frac{\partial^2 t}{\partial x^2} \cdot \frac{\Delta x}{2} \right) \right\} \Delta \tau$$

$$= \lambda \Delta y \Delta z \frac{\partial^2 t}{\partial x^2} \Delta x \Delta \tau$$

The similar expression formulae of the quantity of heat consumed in the cube along y direction and z direction can be derived. Summing the quantity of heat along three directions gives the total quantity of heat stored in the cube, ΔQ.

$$\Delta Q = \Delta Q_x + \Delta Q_y + \Delta Q_z$$

$$= \lambda \Delta x \Delta y \Delta z \frac{\partial^2 t}{\partial x^2} \Delta \tau + \lambda \Delta x \Delta y \Delta z \frac{\partial^2 t}{\partial y^2} \Delta \tau + \lambda \Delta x \Delta y \Delta z \frac{\partial^2 t}{\partial z^2} \Delta \tau$$

$$= \lambda \Delta x \Delta y \Delta z \left(\frac{\partial^2 t}{\partial x^2} + \frac{\partial^2 t}{\partial y^2} + \frac{\partial^2 t}{\partial z^2} \right) \Delta \tau \tag{5-3}$$

The total quantity of heat stored in the cube makes cube temperature increasing. The increasing quantity of heat in the rock cube can be expressed as follows:

$$\Delta Q = \rho C \Delta x \Delta y \Delta z \Delta t \tag{5-4}$$

where, ρ, C, and Δt stand for the density, specific heat, and temperature increment of the rock cube respectively.

Because the total quantity of heat stored in the cube is equal to the increasing quantity of heat, we can get the following equation from formulae (5-3) and (5-4).

$$\frac{\lambda}{\rho C} \left(\frac{\partial^2 t}{\partial x^2} + \frac{\partial^2 t}{\partial y^2} + \frac{\partial^2 t}{\partial z^2} \right) = \frac{\Delta t}{\Delta \tau}$$

If $\Delta \tau$ is tending to zero, the above equation can be expressed as follows:

$$\frac{\lambda}{\rho C} \left(\frac{\partial^2 t}{\partial x^2} + \frac{\partial^2 t}{\partial y^2} + \frac{\partial^2 t}{\partial z^2} \right) = \frac{\partial t}{\partial \tau} \tag{5-5}$$

Assuming that $a = \frac{\lambda}{\rho C}$ and $\left(\frac{\partial^2}{\partial x^2} + \frac{\partial^2}{\partial y^2} + \frac{\partial^2}{\partial z^2} \right) = \nabla^2$ (called Laplace operator), equation (5-5) is changed into the differential equation of rock heat-transfer,

$$a \nabla^2 t = \frac{\partial t}{\partial \tau} \tag{5-6}$$

where, a, t, and τ mean thermal diffusivity, temperature, and time respectively.

5.8.4 Rock Heat-transfer in High-temperature Non-ferrous Metal Mine Stopes Due to Spontaneous Combustion of Ore

Fig. 5-21 shows heat-transfer in high-temperature mine stopes due to spontaneous combustion of ore. Supposing that the quantity of heat generated by spontaneous combustion of ore in high-temperature mine stopes keeps constant, surrounding rock surface can be considered as a half-infinite nonstable one-dimension heat-transfer problem. Its differential equation is as follows:

$$\frac{\partial t(x,\tau)}{\partial \tau} = a\frac{\partial^2 t(x,\tau)}{\partial x^2} \quad (\tau > 0, 0 < x < \infty) \tag{5-7}$$

Before spontaneous combustion of ore, all points in ore have the same temperature T_0. The initial condition is that $t(x,0) = T_0$ when $\tau = 0$. If the gas being at temperature T_c flows over the surrounding rock surface and makes the surface temperature keep unchanged, the boundary condition

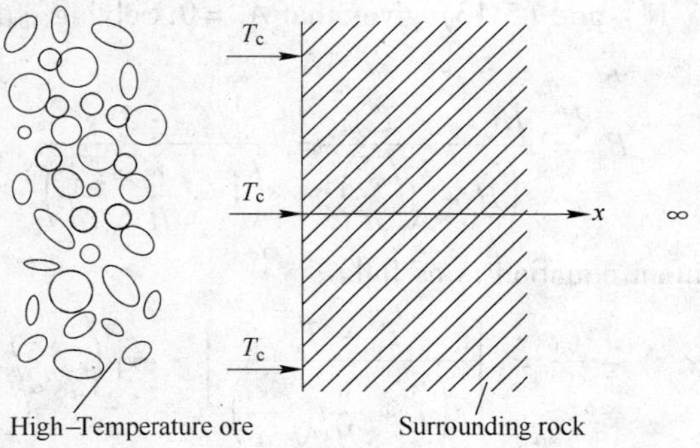

Fig. 5-21 Heat-transfer in high-temperature mine stopes due to spontaneous combustion of ore

can be expressed as follows:

$$\left.\begin{aligned} & t(0,\tau) = T_0 \\ & t(\infty,\tau) = \text{constant} \\ & \frac{\partial t(\infty,\tau)}{\partial x} = 0 \end{aligned}\right\}(\text{the first kind of boundary condition}) \tag{5-8}$$

If there exists an interchange between the rock surface and the flowing gas, the boundary condition should be expressed as follows:

$$\left.\begin{aligned} & -\frac{\partial t(0,\tau)}{\partial x} + H[t(0,\tau) - T_c] = 0 \\ & \frac{\partial t(\infty,\tau)}{\partial x} = 0 \end{aligned}\right\}(\text{the third kind of boundary condition}) \tag{5-9}$$

in which, $H = \frac{h}{\lambda}$ where h and λ stand for heat exchange coefficient and thermal conductivity respectively. By using Laplace transform, differential equation (5-7) is transformed into the following equation

$$t''(x,s) - \frac{s}{a}t(x,s) + \frac{T_0}{a} = 0 \tag{5-10}$$

The general solution of equation (5-10) is as follows:

$$t(x,s) - \frac{T_0}{s} = A\text{ch}\sqrt{\frac{s}{a}}x + B\text{sh}\sqrt{\frac{s}{a}}x = A_1 e^{\sqrt{\frac{s}{a}}x} + B_1 e^{-\sqrt{\frac{s}{a}}x} \tag{5-11}$$

where, $\text{ch}\sqrt{\frac{s}{a}}x = \frac{e^{\sqrt{\frac{s}{a}}x} + e^{-\sqrt{\frac{s}{a}}x}}{2}$, $\text{sh}\sqrt{\frac{s}{a}}x = \frac{e^{\sqrt{\frac{s}{a}}x} - e^{-\sqrt{\frac{s}{a}}x}}{2}$, $A_1 = \frac{A+B}{2}$, and $B_1 = \frac{A-B}{2}$.

If there exists the third kind of boundary condition, formula (5-9) can be transformed by Laplace transform into formulae (5-12) and (5-13).

$$-t'(0,s) + H\left[t(0,s) - \frac{T_c}{s}\right] = 0 \qquad (5\text{-}12)$$

$$t'(\infty, s) = 0 \qquad (5\text{-}13)$$

Solving equations (5-11) and (5-13) gives that $A_1 = 0$. Solving equations (5-11) and (5-12) gets that

$$B_1 = \frac{H(T_0 - T_c)}{s\left(H + \sqrt{\frac{s}{a}}\right)} = \frac{T_0 - T_c}{s\left(1 + \frac{1}{H}\sqrt{\frac{s}{a}}\right)}$$

The solution of phantom equation is as follows:

$$t(x,s) - \frac{T_0}{s} = \left[\frac{T_0 - T_c}{s\left(1 + \frac{1}{H}\sqrt{\frac{s}{a}}\right)}\right] \cdot \exp\left(-\sqrt{\frac{s}{a}}x\right) \qquad (5\text{-}14)$$

Because $L\left[\frac{b}{s(b+\sqrt{s})}\exp(-k\sqrt{s})\right] = \text{erfc}\frac{k}{2\sqrt{\tau}} - e^{-bk}e^{b^2\tau}\text{erfc}\left(b\sqrt{\tau} + \frac{k}{2\sqrt{\tau}}\right)$ and $L\left[\frac{1}{\left(1+\sqrt{\frac{s}{b}}\right)}e^{-k\sqrt{s}}\right] = \sqrt{\frac{b}{\pi\tau}}e^{-\frac{k^2}{4\tau}} - be^{k\sqrt{b}+b\tau}\text{erfc}\left(\sqrt{b\tau} + \frac{k}{2\sqrt{\tau}}\right)$, equation (5-14) can be transformed by using Laplace inverse transform into the following equation, i. e. the solution of the problem:

$$t(x,\tau) = T_0 + (T_0 - T_c)\left[\text{erfc}\frac{x}{2\sqrt{a\tau}} - \exp(Hx + H^2 a\tau) \cdot \text{erfc}\left(\frac{x}{2\sqrt{a\tau}} + H\sqrt{a\tau}\right)\right] \qquad (5\text{-}15)$$

where, error function $\text{erf}(x) = \frac{2}{\sqrt{\pi}}\int_0^x e^{-t^2}dt$ and complementary error function $\text{erfc}(x) = 1 - \text{erf}(x) = \frac{2}{\sqrt{\pi}}\int_x^\infty e^{-t^2}dt$.

Equation (5-15) is the temperature distribution function in the surrounding rock of high-temperature mine stopes at any time τ under the condition of that the quantity of heat generated by spontaneous combustion of ore keeps constant and there exists an interchange between the rock surface and the hot air in the stopes.

5.9 3D Simulating Analysis of a High-temperature Non-ferrous Metal Mine Stope Reinforced in Advance with Grouted Cable Bolts by Using ANSYS Software

5.9.1 Introduction

Although the anchoring technique of fully grouted cable bolts finds a wide application in the reinforcement of slopes, tunnels, deep foundation pit, underground caverns and mine stopes, the rock anchoring mechanism of fully grouted cable bolts in non-ferrous high-temperature metal mine stopes has not been explored till now. Influence of high temperature on surrounding rock and its anchoring system can not be neglected due to the existence of high-temperature mine stopes produced by spontaneous combustion of high-sulphur ore in some non-ferrous metal mines. High-temperature mine stopes are formed due to the oxidation and spontaneous combustion of colloidal sulphur pyrite ore in the process of mining of Dongxiang Copper Mine in China. The embedded depth, horizontal section width along inclined direction, strike length, vertical height, and pitch angle of the studied high-temperature mine stope are 89. 75m, 12. 5m, 7m, 32. 5m, and 30° respectively. The surrounding rock of the stope is granite porphyry. By using advanced engineering simulating analysis software ANSYS, the 3D mine stope temperature model based on the in-situ experiment of reinforcement with grouted cable bolts in the high-temperature stope hangingwall of Dongxiang Copper Mine is established and analyzed. A comprehensive rock anchoring theory of grouted cable bolts in non-ferrous high-temperature metal mine stopes, i. e. theory of swelling-rising concrete arch with tree-root-shaped cable bolt reinforcement, is presented. Having no disadvantages of each other kind anchoring theory which considers only a single effect, it expounds briefly and explicitly the rock anchoring mechanism of grouted cable bolts in non-ferrous high-temperature metal mine stopes. The achievements are helpful and beneficial to underground deep mining, mine extraction at the risk of spontaneous combustion of ore, and the design and construction of rock engineering projects which need anchoring reinforcement, such as slopes, tunnels, caverns, and mine stopes in high-temperature environment or the environment affected probably by high temperature.

5.9.2 Simulating Limits

The horizontal projection length of the stope along inclined direction is 68. 79m owing to the pitch angle 30°. Three times horizontal section width of the stope is added to the left and right sides in the horizontal direction of the stope individually. So the total simulating width is 143. 79m. The embedded depth plus two times vertical height equals the total simulating height 154. 75m. Three times strike length of the stope gives the total simulating strike length 21m. Fig. 5-22 is the schematic section of simulating stope limits.

5.9.3 Model Selection

3D simulating ANSYS analysis of high-temperature mine stopes reinforced with fully grouted cable bolts is a thermal-stress coupled problem. Thermal-structural coupled solid element SOLID98 (a tetrahedron with 10 nodes) and direct coupled method are used in order to solve the problem. There are three rock types: granite porphyry, rock reinforced with cable bolts, and filling medium. Three normal-temperature stope models are full-fill, half-fill, and no-fill. Normal temperature is 25℃. High-temperature stope models have no filling medium. Twelve high-temperature stope models are 35℃, 60℃, 100℃, 200℃, 300℃, 400℃, 500℃, 600℃, 700℃, 800℃, 900℃, and 1000℃.

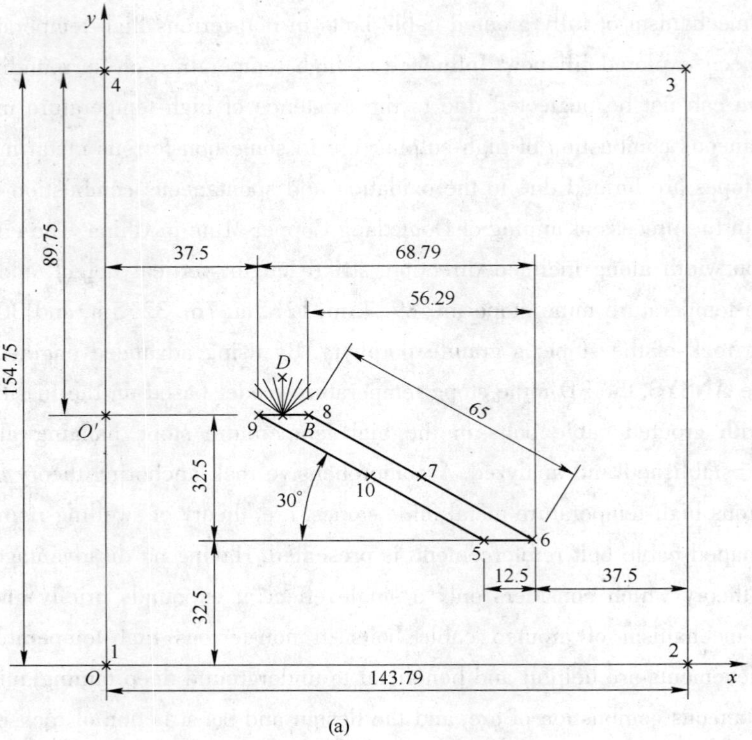

(a)

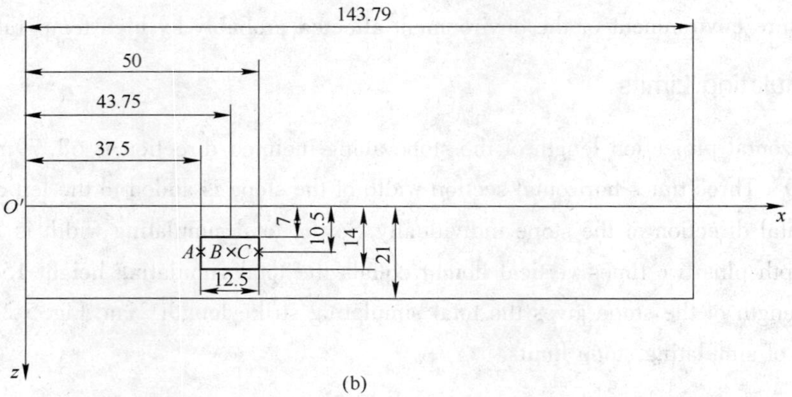

(b)

Fig. 5-22 Schematic section of simulating limits (unit: m)

(a) Vertical section along inclined direction through the rear end surface of the limits;

(b) Horizontal section through the bottom surface of the stope hangingwall

5.9.4 Mechanical Parameters of ANSYS Analysis

The elastic modulus of granite porphyry at normal temperature is relatively great, but it decreases rapidly with the rise of temperature. In the process of ANSYS analysis, influence of earth temperature gradient on the in-situ rock stress is taken into account in normal-temperature (25℃)

models and is not considered in all high-temperature models. Table 5-4 shows mechanical parameters of ANSYS analysis of normal-temperature and high-temperature models. Note that for the datum items with parentheses in Table 5-4, the data out of and in parentheses belong to normal-temperature and high-temperature models respectively. For the datum items with no parentheses, the data belong to both normal-temperature and high-temperature models.

Table 5-4 Mechanical parameters of ANSYS analysis of normal-temperature and high-temperature models

Parameters	Granite porphyry	Rock reinforced with cable bolts	Filling medium
Thermal conductivity KXX, W/(m · ℃)	0.11	0.21	0.16
Elastic modulus EX, Pa	15.3E009(6E009)	16.8E009(12E009)	2.48E009
Poisson' s ratio PRXY	0.3	0.25	0.18
Linear expansion coefficient ALPX, ℃$^{-1}$	1.0E-005	1.1E-005	1.16E-005
Density DENS, kg/m^3	3340(2872)	3344(2876)	2080
Specific heat C, J/(kg · ℃)	2.4E-004	2.1E-004	4.3E-004

5.9.5 Establishing Models

Assuming that XOY surface of Cartesian Coordinate System is on the rear end surface of the model along strike direction and the coordinate origin is located at the left-bottom point on that rear end surface. The positive X direction is horizontal and to the right. The positive Y direction is vertical and up. First of all, we create 10 key points 1(0, 0, 0), 2(143.79, 0, 0), 3(143.79, 154.75, 0), 4(0, 154.75, 0), 5(93.79, 32.5, 0), 6(106.29, 32.5, 0), 7(78.145, 48.75, 0), 8(50, 65, 0), 9(37.5, 65, 0), and 10(65.645, 48.75, 0) (see Fig. 5-22(a)). Secondly lines are created by using these key points. A circle is obtained by using the center of circle (43.75, 65, 0) and the point in circumference (54.25, 65, 0). Deleting the lower half of the circle gives the circular arc of rock reinforced with cable bolts. Then surfaces are created by using the lines. The surfaces are overlapped and glued. Finally 3D stope models are formed from the surfaces extruded along normal direction. Filling medium consists of two parts the upper half and the lower half. Deleting the upper half gets the half-fill model and deleting two halves gets the no-fill model.

5.9.6 Meshing Models

The lines are manually divided into different line segment numbers. The lines passing through the whole simulating width, height, and depth limits are divided into eight, eight, and four segments respectively. The other lines are divided into two, three, or four segments. Different solid models are meshed with corresponding materials. Fig. 5-23 shows the meshed elements of the 3D half-fill stope model.

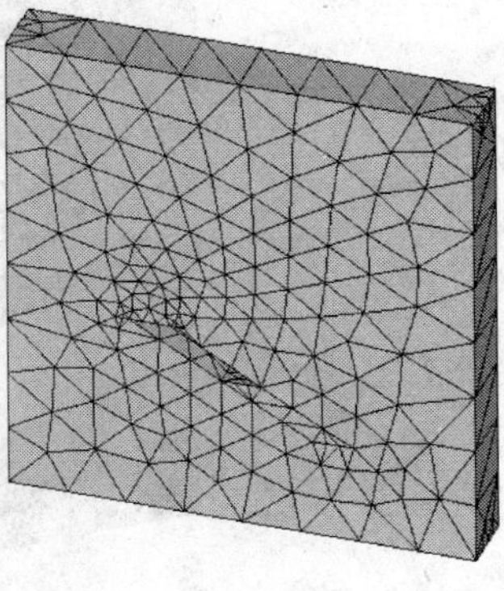

Fig. 5-23 Meshed elements of the 3D half-fill stope model

5.9.7 Loading

Loading of normal-temperature models is steady-state. Length unit is SI (MKS) and temperature unit is Celsius. The initial uniform temperature is 25℃. Inertia load in Y direction equals acceleration of gravity $9.8\mathrm{m/s^2}$. The boundary condition of displacement is that $U_x = U_z = 0$ for all nodes on the two end surfaces of the model along inclined direction, $U_y = U_z = 0$ for all nodes on the bottom end surface of the model along vertical or height direction, and $U_z = 0$ for all nodes on the two end side surfaces of the model along strike direction.

Loading of high-temperature models is transient and has 6 loading sub-steps. According to twelve different high-temperature models, the corresponding temperature loads are respectively exerted on all nodes on the two end side surfaces of the mined-out space along inclined direction and on the top and bottom surfaces of the space. Length unit, temperature unit, the initial uniform temperature, inertia load, and the boundary condition of displacement of high-temperature models are the same as those of normal-temperature models.

5.9.8 Solving Results

Path *ABC* along width direction on the stope hangingwall rock surface, and path *BD* along height direction in the hangingwall rock reinforced with cable bolts are analyzed. The coordinates of points *A*, *B*, *C*, and *D* are *A*(37.5, 65, 10.5), *B*(43.75, 65, 10.5), *C*(50, 65, 10.5), and *D* (43.75, 75.5, 10.5) (see Fig. 5-22). Solving the model by ANSYS software, we achieve the isothermal lines, deformation, displacement in y direction (U_y), the first primary stress (σ_1), the second primary stress (σ_2), and the third primary stress (σ_3) of the surrounding rock of all normal-temperature and high-temperature models. U_y, σ_1, σ_2, and σ_3 curves on paths *ABC* and *BD* are also obtained. Fig. 5-24 shows the typical isothermal lines of high-temperature mine stopes. Fig. 5-25 gives U_y curve on path *ABC* of no-fill model at 600℃. Notice that in Figure 4 the ab-

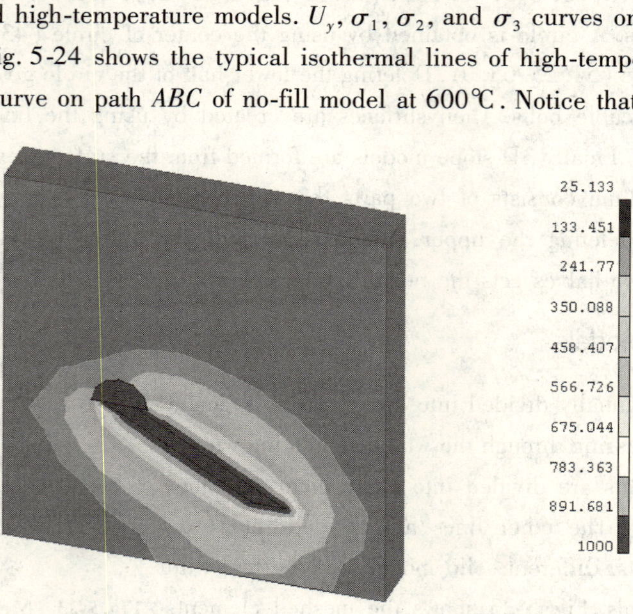

Fig. 5-24 Isothermal lines of high-temperature mine stopes (unit: ℃)

scissa is the distance (m) from point A to any point on path ABC and the ordinate is U_y (m). The other graphs and curves are not listed here due to limited space. The maximum ground surface displacement U_{ysmax}, U_y, σ_1, and σ_3 at points A, B, and C on path ABC, and the maximum and minimum values of U_y on path ABC, i. e. max and min, are listed in Table 5-5. Notice that in Table 5-5 negative and positive displacements stand for subsidence and rising ones respectively, negative and positive stresses means compressive and tensile ones respectively, and the units of displacements and stresses are mm and MPa respectively.

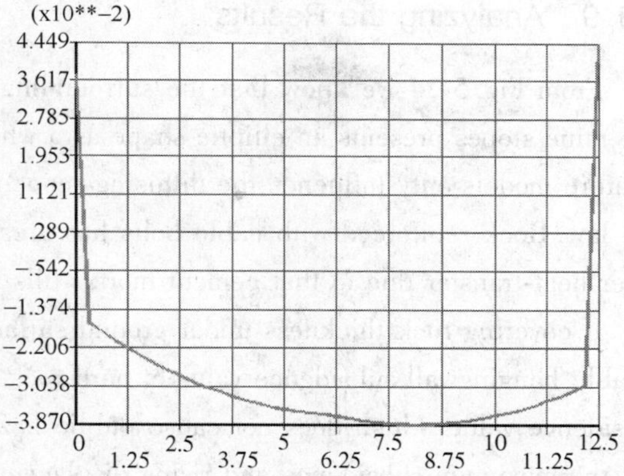

Fig. 5-25 U_y curve on path ABC of no-fill model at 600℃ (unit: m)

Table 5-5 Values of U_y, σ_1, and σ_3 on path ABC and U_{ysmax} of the models

Models	U_{ysmax}	U_y					σ_1			σ_3		
		A	B	C	Max	Min	A	B	C	A	B	C
Full-fill	-37	-11	-25	-29	-11	-29	0.46	0.25	0.05	-1.87	0.03	-0.21
Half-fill	-42	-11	-28	-30	-11	-31	0.57	0.30	0.01	-1.98	0.05	-0.25
No-fill	-48	-11	-29	-31	-11	-32	0.59	0.30	0.01	-2.06	0.06	-0.25
35℃	-42	4	-27	-31	4	-33	2.75	-0.01	0.08	-1.47	-5.12	-1.47
60℃	-17	6	-27	-27	6	-33	2.61	0.01	0.03	-2.39	-8.48	-2.49
100℃	36	8	-27	-25	8	-33	2.42	-0.03	1.25	-4.25	-13.97	-5.10
200℃	55	14	-30	-9	14	-32	2.07	-0.10	1.93	-9.26	-27.33	-10.22
300℃	58	19	-31	3	19	-33	1.89	-0.17	3.13	-14.28	-31.80	-15.25
400℃	69	25	-32	1	25	-35	1.38	-0.43	4.56	-19.31	-38.79	-19.31
500℃	82	31	-34	8	31	-37	1.26	-0.56	5.73	-24 .33	-41.40	-24.33
600℃	90	36	-35	44	44	-38	0.97	-0.33	6.68	-29.36	-81.21	-31.43
700℃	105	43	-37	57	57	-40	0.63	-0.39	7.87	-34.39	-94.68	-36.42
800℃	123	49	-38	35	49	-42	0.43	-1.20	14.32	-39.19	-109.22	-39.19
900℃	135	54	-40	56	56	43	-0.33	-0.50	10.24	-44.44	-121.62	-48.2
1000℃	152	58	41	55	58	-45	-0.49	-1.53	18.06	-49.06	-136.44	-49.06

5.9.9 Analyzing the Results

From Fig. 5-24 we know that the surrounding rock temperature distribution of high-temperature mine stopes presents an elliptic shape as a whole. The temperature sizes of different high-temperature models only influence the diffusing scope of temperature and do not influence its distribution law. Rock reinforced with cable bolts has greater density, less thermal contact resistance, and faster heat-transfer due to that cement mortar fills faults, fissures, and cleavages.

If covering rock thickness under ground surface is between 50m and 500m, the maximum allowable hangingwall subsidence value of hard rock is 20 to 60mm and the allowable ground surface subsidence value, which does not cause public hazards, is 200 to 300mm. From Table 5-5 we see that the maximum subsidence and rising displacements of ground surface are −48mm and 152mm respectively. The maximum hangingwall subsidence and rising displacements are −45mm and 58mm respectively. Both displacements of ground surface and hangingwall are in their allowable scopes.

The surrounding rock rising displacement caused by heated rock expansion can partly or fully offset the subsidence one caused by gravity. Subsidence decrease or even rising displacement of hangingwall is beneficial to its stability. Hangingwall rising displacement occurs if stope temperature is equal to or greater than 100℃. Fig. 5-25 and Table 5-5 show that rising displacements appear at the corner points A and C of hangingwall when stope temperature increases. The central point B has only subsidence displacement and no rising one. From Table 5-5 we understand that when stope temperature increases the first primary stress of point A changes from the tensile to the compressive and both compressive stress of point A and tensile stress of point C increase gradually. Point A is mainly compressed, point C has relatively great tensile stress, and subsidence of point B is relatively great. Partial collapse of points B and C is easy to occur.

The maximum tensile stress (7.87MPa) at 700℃ starts to go beyond the allowable tensile stress (6.81MPa) of the anchored rock, and the maximum compressive stress (81.21MPa) at 600℃ begins to exceed the compressive strength (30MPa) of the surrounding rock at the same temperature. Mining is safe if stope temperature is equal to or less than 500℃. For actual temperature of high-temperature stopes in the mine is generally between 35℃ and 300℃, the stope hangingwall anchoring parameters can meet the requirements of surrounding rock stability in the process of extraction.

5.10 Rock Anchoring Mechanism in High-temperature Non-ferrous Metal Mine Stopes——Theory of Swelling-rising Concrete Arch with Tree-root-shaped Cable Bolt Reinforcement

5.10.1 Model of "Theory of Swelling-rising Concrete Arch with Tree-root-shaped Cable Bolt Reinforcement"

In terms of anchoring rock with a single fully grouted cable bolt, it looks like the root of a tree called "the tree root of a cable bolt". The cable bolt bound up with cement mortar, grouting in cracks such as faults and soft layers, and grouting in fissures and cleavages are similar to the main trunk, the primary boughs, and the secondary twigs of the tree root respectively. The construction process of fully grouted cable bolts is just the growing one of "the tree roots of cable bolts".

The anchoring rock of a single fan-shaped row of fully grouted cable bolts is similar to a reinforced concrete arch bridge. The steel bar in the bridge is called "tree-root-shaped cable bolt reinforcement" (See Fig. 5-26). The fractured rock bound up with cement mortar corresponds to concrete, and therefore the anchoring rock is referred to as "concrete arch with tree-root-shaped cable bolt reinforcement" (See Fig. 5-27). An arch has two main characteristics: one is that an arch mainly bears compressive force and the other is that there is great horizontal thrust at the foot of an arch. "Concrete arch with tree-root-shaped cable bolt reinforcement" takes advantage of great compressive strength of rock completely and transmits the loads of stope hangingwall to surrounding rock. Stress distribution of "concrete arch with tree-root-shaped cable bolt reinforcement" is more reasonable than that of beams.

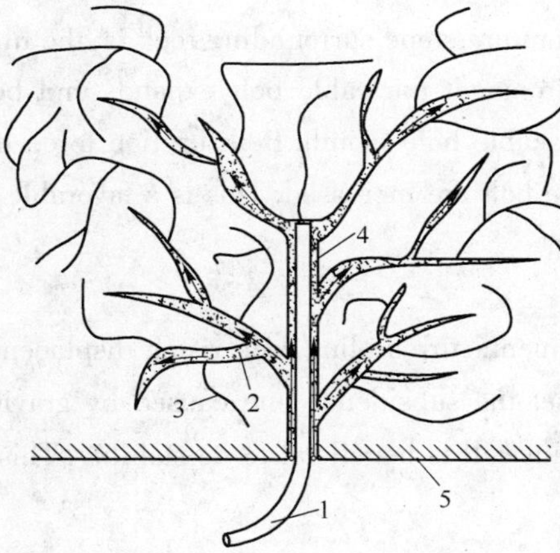

Fig. 5-26 Diagram of "tree-root-shaped cable bolt reinforcement"

1—Cable bolts; 2—Faults; 3—Fissures; 4—Mortar; 5—Surrounding rock

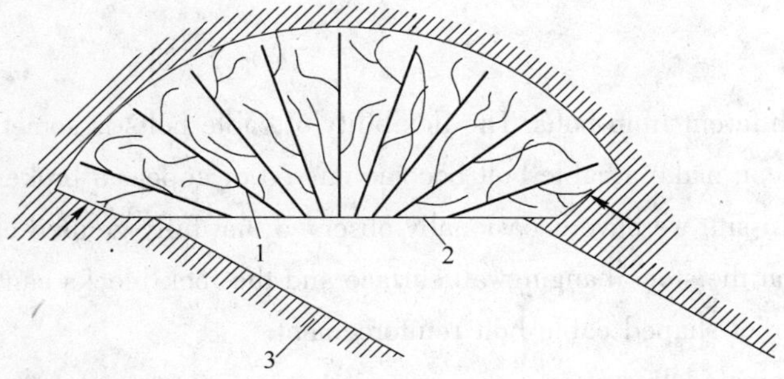

Fig. 5-27 Diagram of concrete arch with tree-root-shaped cable bolt reinforcement

1—Tree-root-shaped cable bolt reinforcement; 2—Concrete arch with tree-root-shaped cable bolt reinforcement; 3—Surrounding rock

5.10.2 Main Viewpoints of "Theory of Swelling-rising Concrete Arch with Tree-root-shaped Cable Bolt Reinforcement"

"Theory of swelling-rising concrete arch with tree-root-shaped cable bolt reinforcement" includes two main viewpoints, i. e. swelling-rising effect and effect of concrete arch with tree-root-shaped cable bolt reinforcement.

5.10.2.1 Swelling-Rising Effect

1. Swelling Effect

Surrounding rock of underground high-temperature stopes is infinite or half-infinite for embedded depth of the stopes is relatively great. The thermal stress called swelling ground pressure is generated in rock due to the fact that energy of heated rock expansion can not be released. This is an unfavorable aspect of high-temperature swelling effect on anchoring rock.

The closer to the high-temperature stope surrounding rock is, the higher the temperature in the rock is. In high-temperature environment the cable bolt expands and becomes wedge-shaped. The larger end of the wedge is at the cable hole mouth. Both friction force between cable and grouting and cohesive strength of the cable bolt are increased. This is a favorable aspect of high-temperature swelling effect on anchoring rock.

2. Rising Effect

In high-temperature environment, surrounding rock rising displacement caused by heated rock expansion can partly or fully offset the subsidence one caused by gravity. Subsidence decrease or even rising displacement of hangingwall is beneficial to its stability. This is high-temperature rising effect on anchoring rock.

5.10.2.2 Effect of Concrete Arch with Tree-Root-Shaped Cable Bolt Reinforcement

Effect of concrete arch with tree-root-shaped cable bolt reinforcement includes the following five main effects.

1. Hanging Effect

Cable bolts are different from bolts. The flexibility of cable bolts is something like that of tree roots. Action of faults can make a cable bolt become curved or appear a broken line, but its anchoring force still exists. In-situ we have occasionally observed that fully grouted cable bolts hang some unstable rock blocks on the stope hangingwall surface and the rock blocks can not drop. This is the hanging effect of tree-root-shaped cable bolt reinforcement.

2. Effect of a Laminated Beam

Tree-root-shaped cable bolt reinforcement, which is something like linen thread in the sole of cloth shoes, sews up the rock layers of stope hangingwall together to form a laminated component as a whole. Resisting force between rock layers increases greatly, and the bending strength, shear strength, and rigidity of hangingwall rock also grow.

3. Effect of Injection Anchorage

That cement mortar injects and fills faults, fissures, and cleavages makes rock being in a 1D or 2D stress state become in a more stable 3D one. Injection can also increase the strength and deformation modulus of soft constructional surfaces in surrounding rock. Strip and pulse-shaped grouting, which is something like the root hair of a tree, wraps hangingwall fissured rock into a wedge-shaped thing. This wedge wedges the top part of the stope. The space system of forces of the whole stope is balanced, and therefore no hangingwall collapse occurs.

4. Waterproof and Impervious Effect

Injection fills faults, fissures, and cleavages in surrounding rock to form an artificial impervious hangingwall with certain thickness. Influence of water on surrounding rock is alleviated or effectively protected. This waterproof and impervious effect is the same as principles of water-protection, sand-fixation, and water and soil conservation of tree roots.

5. Effect of Anchoring in Advance

Fully grouted cable bolts reinforce stope hangingwall in advance before in-situ rock is disturbed. The strength of surrounding rock itself is maintained and acquires further improvement and enhancement owning to the fact that tree-root-shaped cable bolt reinforcement restrains deformation of surrounding rock. It is protected that a 3D stress state of surrounding rock is transformed into a plane stress state in the process of mine extraction. The stability of high-temperature stopes is guaranteed.

Professional Terms

rock mechanics 岩石力学
classification of rock 岩石分类
stress 应力
strain 应变
deformation (displacement) 变形
in-situ stress 原岩应力
tensile stress 拉应力
compressive stress 压应力
mechanical property test of rock 岩石力学性质试验
strength 强度
stiffness 刚度
stability 稳定性
tensile strength 抗拉强度
compressive strength 抗压强度
uniaxial compressive strength 单轴抗压强度
stress-strain curve 应力-应变曲线
stress concentration 应力集中
Young's modulus 杨氏模量(弹性模量)
Poisson's ratio 泊松比
rock specimen 岩石试样,岩石标本
field test 现场试验
high-temperature mine stope 高温采场
surrounding rock 围岩
rock heat-transfer model 岩体传热模型
simulating analysis 仿真分析
rock anchoring mechanism 岩体锚固机理
Theory of Swelling-Rising Concrete Arch with Tree-Root-Shaped Cable Bolt Reinforcement 膨胀抬升锚索树根筋混凝土拱理论(简称

胀锚拱理论）

swelling-rising effect 膨胀抬升作用

effect of concrete arch with tree-root-shaped cable bolt reinforcement 锚索树根筋混凝土拱作用

Exercises

5-1 解释“size effect”的含义。

5-2 胀锚拱理论的要点是什么？

5-3 psi, b, lb 各代表什么单位？

6 Mining Methods of Non-ferrous Metal Deposits

本章介绍有色金属矿床采矿方法，包括采矿方法选择、地下采矿方法（如留矿采矿法、充填采矿法和分段崩落采矿法）、露天开采方法和设备（如露天钻机、前端装载机和铲运机等），最后介绍一个露天铜矿开采实例。

6.1 Mining Method Selection

6.1.1 Classification of Deposits and Rock

After a mineral deposit has been discovered, delineated and evaluated, the next step is to select a mining method that is physically, economically and environmentally adaptable to recovering the mineral from the deposit. The term "mineral deposit" is used to denote a concentration of mineral (including rock and coal). Even where there is considerable past experience available, many of the critical design criteria must be determined utilizing only information obtained from an examination of exploratory drill core specimens.

The factors that affect this selection are as follows: (1) the spatial characteristics of the deposit (size, shape, attitude and depth); (2) the physical (or mechanical) properties of the mineral deposit and surrounding rock; (3) ground water and hydraulic conditions; (4) economic factors, including grade of the ore, comparative mining costs and desired production rates; and (5) environmental factors, such as the preservation of the surface overlying the mine, and the prevention of air and water pollutants. Of these factors, the spatial characteristics of the deposit and the physical (or mechanical) properties of the mineral and surrounding rock limit the methods that can be employed to mine it.

Six types of mineral deposits are recognized, as follows:

(1) Massive——A deposit of considerable lateral and vertical extent in which mineralization (or ore) is relatively uniformly distributed. Disseminated copper ores are typical deposits. Dome

salt also is included in this classification.

(2) Bedded or Tabular——A mineral deposit that parallels the stratification, most often in sedimentary rock, usually laterally extensive, and of limited thickness. Most coal measures and some evaporite mineral deposits (e. g. , potash) fall in this class.

(3) Narrow Vein——A zone or belt of mineralization (ore) typically long, narrow (less than 3m), often dipping steeply, and usually lying within boundaries separating it from neighboring rock. Many gold and other metallic minerals occur in narrow veins.

(4) Wide Vein——As defined under narrow vein except that the vein thickness is greater than 3m.

(5) Lenticular or Pocket——An isolated ore body, or an enrichment of limited vertical and horizontal extent in a massive, bedded or vein deposit. Lead, zinc and iron ores often occur in this type of deposit.

(6) Placer——A surface or near-surface deposit, usually tabular and may be of considerable areal extent, containing mineral particles (e. g. , gold, platinum) in detritus (e. g. , sand, gravel).

The rock materials associated with mineral deposits are classified as follows:

(1) Detritus——Sediments and other accumulations of solid particles produced by the mechanical and/or chemical disintegration of rocks, and usually having a low cohesive strength and very low compressive strength (e. g. , soil, sand, gravel, alluvium, playa).

(2) Jointed and Fractured——Unbonded—A rock mass, often occurring on or near the surface or in fault zones, in which the joints or fractures contain alteration or decomposition products, such as clay materials, and across which the cohesive strength is very low. Such a rock mass can be considered as particulate, but with a particle size greater than that for soil.

(3) Jointed and Fractured, with Partial Bonding——A rock mass in which the joints or fractures are partially recemented, producing a rock material with a low-to-medium cohesive strength.

(4) Laminar Thin-Bedded——A rock mass, generally of sedimentary origin, in which the laminae thicknesses average less than 0. 3m. The partings between laminae may be partially bonded and the rock mass may be fractured. The bedded shales and sandstones found in coal-mine roofs are typical of this class of rock material.

(5) Laminar Thick-Bedded——A rock mass in which the average lamina thickness is greater than 0. 3m. Some bond may occur across partings and the rock mass may be fractured.

(6) Massive——A relatively unjointed or unfractured rock mass, or a jointed or fractured rock mass in which the degree of bonding across joints or fractures produces a strong rock material. Recemented breccias and conglomerates are included in this rock class.

In the following subsections the various mining methods are briefly described, and the spatial characteristics of the deposit and physical properties of the minerals and surrounding rock in which each method can be employed are specified. Marine mining is not considered in this section.

6.1.2 Selection of Underground Mining Methods

If the depth of an ore deposit is such that removal of the overburden makes surface mining unprofitable, underground methods should be considered. The problem of recovering the mineral from such a deposit is reduced to selecting or developing a mining system that will exclude other options on a safety and profit basis and at the same time provide adequate ground support to protect working areas and, in some instances, to preserve the surface. Because ground support is a necessary element in this process, the mining systems listed in Table 6-1 and described in the following subsections are classified on this basis. Also, because ground support is so dependent on the spatial characteristics of the deposit and mechanical properties of the mineral and surrounding rock materials, each description includes a specification of the deposit characteristics in which the method is applicable. Three broad classes of mining methods are recognized as follows:

(1) Methods in which the underground openings (rooms or stopes) created by the extraction of the mineral are self-supported in that no regular artificial method of support is employed: that is, openings in which the loads due to the weight of the overburden or tectonic forces are carried on the sidewalls and/or pillars of unexcavated mineral of rock. This specification does not preclude the use of rockbolts or other light systems of support, provided that this artificial support does not significantly affect the load carried on the self-supported structure.

(2) Methods in which stopes or rooms require significant support, that is, support to the degree that a part of the superincumbent load is carried on the support system.

(3) Methods in which, because of the spatial and mechanical properties, the deposit is induced to cave under the action of gravity to produce better results than more selective methods.

Table 6-1 Underground mining methods

I	Self-supporting openings	Open-stope mining	Isolated openings	
			Pillared open stopes	Open stoping with random pillars
				Open stoping with regular pillars
		Room-and-pillar mining		
		Sublevel stoping		
		Shrinkage stoping		
		Stull stoping		
Ⅱ	Supported openings	Cut-and-fill stoping		
		Square-set-and-fill stoping		
		Longwall mining		
		Shortwall mining		
		Top slicing		
Ⅲ	Caving methods	Sublevel caving		
		Block and panel caving		

This classification of underground mining methods is essentially the same as that given by Jackson and Gardner. Similar classifications are given by Morrison and Lewis and Clark. Because pillar removal and pillar robbing are secondary mining operations, they are not considered in this section.

Often there is no precise division between requirements for these classes of mining methods. In fact, both open stoping and block caving have been employed in the same mineral deposit. Such a situation might result from a change in properties of the rock materials within or surrounding the deposit, or a change in spatial characteristics of the deposit, such as a thinning out, or the technological or economic changes that make one method preferable to another. However, the choice of a mining method usually is dictated more by the spatial or mechanical characteristics of the deposit than by other factors, and sometimes uniquely.

The principal environmental problems created by underground mining are the discharge of acid mine water into streams, and surface scars produced by subsidence, both concurrent and subsequent to mining. In the United States, the discharge of acid mine water is a problem generally associated with coal mining in the eastern states. Surface subsidence is common in the mining of massive low-grade deposits by caving methods. However, many underground mines, especially those in bedded and large lenticular deposits, and from which a high extraction has been obtained, ultimately subside, often to the degree that utilization of the overlying surface is placed in permanent jeopardy. Coal and evaporite mineral mines are typical examples. Generally, an underground mine does not involve broad environmental problems other than subsidence, although it may present one or more occupational problems.

6.1.2.1 Self-Supported Openings

Self-supported openings—that is, openings in which the superincumbent loads are carried on the sidewalls or pillars of unexcavated rock or mineral—can be mined in any type of mineral deposit except placer. However, the size of opening that can be excavated will depend on the type of rock materials that comprise the sidewalls and pillars. For example, the span (minimum wall-to-wall, wall-to-pillar or pillar-to-pillar dimension) that will stand unsupported may range from virtually zero for closely jointed or thinly laminated rock materials in which there is no cohesive strength across joints or partings, to more than 30m for massive bodies of rock. Thus, in general, the size of a self-supported opening will depend on the spacing and strength across mechanical defects in the rock material and on the depth and orientation of the opening.

Mining methods employing self-supported openings can be divided into two broad classes—open-stope mining and room-and-pillar mining.

1. Open-Stope Mining

Strictly defined, an open stope is an underground opening from which a valuable mineral has been removed and in which no timber or other material was used to support the walls or roof. More common usage includes as open stopes those in which walls and/or roof may be supported by pillars of ore or waste, or by stulls, roof bolts or other means. We subscribe to the latter definition and, from a structural standpoint, divide open stopes into two classes: isolated (single) openings and pillared stopes (multiple openings).

Isolated Openings

An isolated opening is an unpillared and otherwise unsupported underground opening that is essentially outside the zone of influence of other underground openings. Isolated pockets, lenses and shoots of ore have been mined in this manner. Also, shafts, development drifts and excavations for civil work projects (tunnels, underground chambers for power stations) may be included in this category.

In general, isolated openings can be mined in any rock type where the physical characteristics permit. The maximum span that can be mined as an isolated opening will depend upon the depth of the deposit and upon the geologic and physical parameters of the rock surrounding the ore. For example, isolated openings with spans up to 30m and at a depth of 91m have been mined in chert breccias; spans from 15 to 23m in fractured jaspilite at depths up to 305m; and spans from 15 to 18m in relatively unfractured dolomitic limestone at depths up to 91m. Fig. 6-1 illustrates an isolated opening.

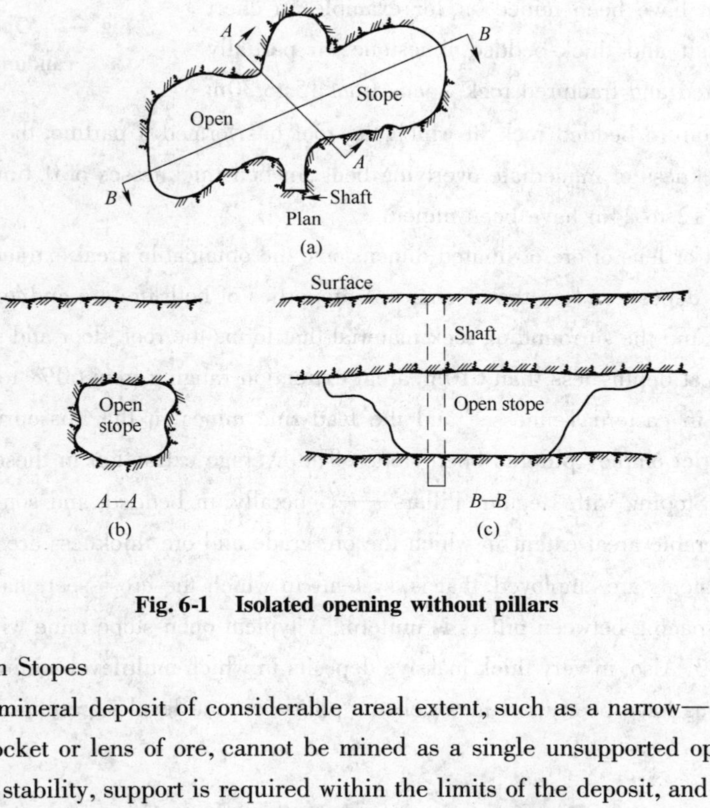

Fig. 6-1 Isolated opening without pillars

Pillared Open Stopes

Generally, a mineral deposit of considerable areal extent, such as a narrow—or wide-vein deposit or a large pocket or lens of ore, cannot be mined as a single unsupported open stope (opening). To maintain stability, support is required within the limits of the deposit, and if this support is effected by leaving areas of unexcavated ore or waste, the system of mining is referred to as pillared open stoping.

From a structural standpoint such a system of stopes corresponds to a system of multiple openings: that is, openings sufficiently close to one another so that the stress distribution around one opening affects the stress distribution around an adjacent opening, and vice versa.

Pillared open stope mining is effected with both random and regular pillar systems.

(1) Open Stoping with Random Pillars——Open stoping with pillars that are randomly spaced and/or random in size is illustrated in Fig. 6-2. This method is used in mining the following:

1) Larger pockets or lenses of ore, especially if the ore grade and/or thickness of the deposit

is variable. Whenever possible, the pillars are left in leaner ore. If ore is high-grade, pillars usually are extracted in final mining.

2) Bedded and narrow or relatively wide vein deposits that dip at any angle (usually less than 45°), that is, at an angle such that the broken ore will not flow under the action of gravity. Also, this method is more likely to be employed when the ore grade or ore thickness is variable.

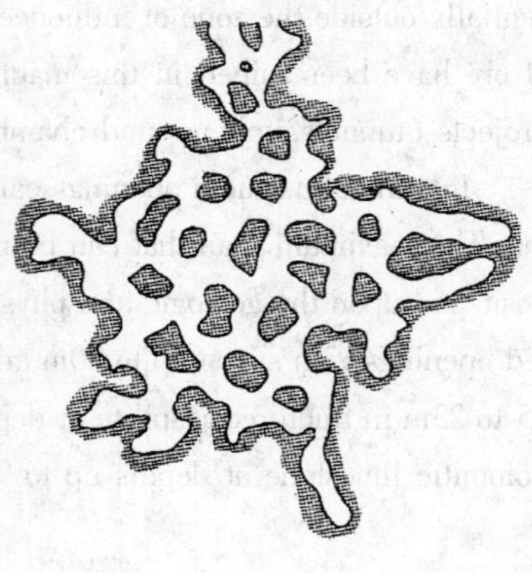

Fig. 6-2 Open stope with random pillars

In randomly pillared openings, the span will depend on the quality of the roof rock. If it is massive, openings with spans up to 30m have been mined as, for example, in chert breccia, dome salt and thick-bedded limestone. In partially recemented jointed and fractured rock, spans from 15 to 30m are not uncommon. In bedded rock in which the roof has formed a parting, the roof span will depend on the thickness of immediate overlying beds. In bed thicknesses of 0.6m or more, openings with spans from 12 to 24m have been mined.

In a pocket or lens of ore of limited dimensions, the obtainable areal extraction will depend on the depth of the deposit and on the mechanical properties of both the ore and/or rock material that form the pillars, and the surrounding rock material that forms the roof, floor and sidewalls of the deposit. Generally, at depths less than 610m, areal extraction ranging from 60% to 80% is obtained. The zinc mines in eastern Tennessee and the lead-zinc mines in the Missouri-Kansas-Oklahoma (Tri-State) District employ pillared open stopes. The average extraction in these mines is 75%.

(2) Open Stoping with Regular Pillars——Generally, in bedded, and sometimes in vein deposits of considerable areal extent in which the ore grade and ore thickness are relatively uniform, regular pillar systems are employed: that is, systems in which the cross-sectional shape and size of pillars and the spacing between pillars is uniform. A typical open-stope mine with regular pillars is shown in Fig. 6-3. Also, in very thick massive deposits in which multilevel mining is employed, regular pillar systems have been used in which the pillars on one level superimpose those of the next

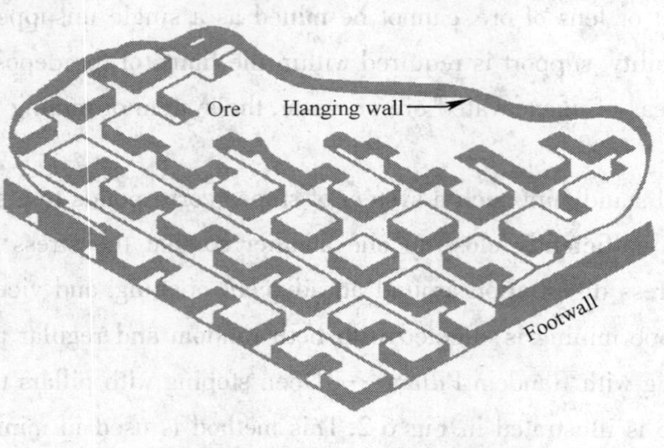

Fig. 6-3 Open stope with regular pillars

lower level. Mining in salt domes follows this procedure as, for example, in the Jefferson Island mine in Louisiana.

In regular-pillared open stopes the span that will stand unsupported depends primarily on the type of roof rock, ranging from 3 to 3.7m for thin-bedded shales to 46m in dome salt. Areal and volume extraction obtained with this type of mining depends on the depth of the deposit and on the mechanical properties of the rock or ore materials that form the pillars. Generally, at depths less than 610m extractions ranging from 60% to 80% are obtained.

2. Room-and-Pillar Mining

The design of room-and-pillar layouts is essentially the same as that for open stoping with regular pillars, except that the former is limited to relatively flat-lying deposits in which the mineral is of comparatively uniform grade and thickness as, for example, coal or evaporite minerals (bed salt, potash, trona and borax). Also, some underground stone quarries employ room-and-pillar systems. This method, illustrated in Fig. 6-4, is effected by mining a grid of rooms separated by pillars of uniform cross section. Many grid layouts have been employed, including systems with rib pillars, and square pillars with checkerboard spacing.

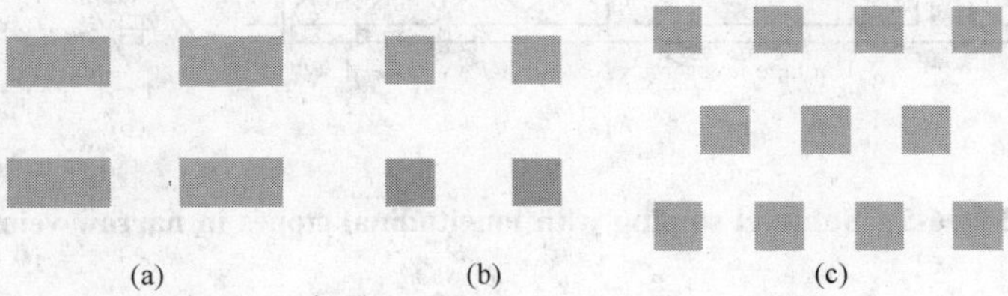

Fig. 6-4 Examples of pillar patterns

(a) Rectangular pillars, regularly spaced; (b) Square pillars, regularly spaced (checkerboard); (c) Staggered checkerboard

The room widths, which usually are made as large as safety will permit, are limited by the characteristics and properties of the immediate roof rock. The cross-sectional size of the pillars, plus the room widths, determine the extraction, and will depend on the depth of the deposit and the strength and other mechanical properties of the rock material that forms the pillars.

In coal mining, room widths range from 4 to 15m, with 9m as an average. Extractions vary from 50% to 70%, with 60% as an average. In evaporite mines, the range of room width is comparable, but extraction ratios generally are higher, ranging from 60% to 90%.

As extreme examples of room-and-pillar mining, in the Saskatchewan Canadian Potash District a 2.3m bed lying at a depth of 945m is mined some distance below an aquifer with a room width of 6.4m and an extraction of 30%. In an experimental oil-shale mine, a 16m section of oil shale lying under 274m of cover was mined with a staggered checkerboard system. The room width was 18m and the pillar cross sections 18m × 18m, giving an extraction of 75%.

3. Sublevel Stoping

Sublevel stoping generally is employed in steeply dipping narrow-and wide-vein and bedded deposits, although this method has been used successfully in relatively flat-lying thick deposits. The deposit thickness may be variable, but the ore grade should be fairly uniform since this method does not lend itself to sorting. The rock material in the hanging and footwall and the ore should be relatively competent—that is, of a type mechanically equivalent to that in which other open-stope methods are employed.

Two basic stope configurations are utilized in sublevel stoping—longitudinal and transverse—and are illustrated in Fig. 6-5 and Fig. 6-6. In both stope configurations the ore is mined from sublevels by benching or ring drilling, and the ore should flow by gravity to the drawpoints.

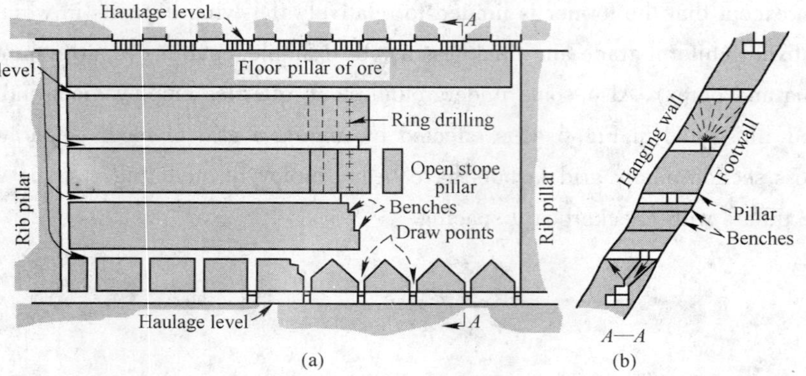

Fig. 6-5 Sublevel stoping with longitudinal stopes in narrow veins

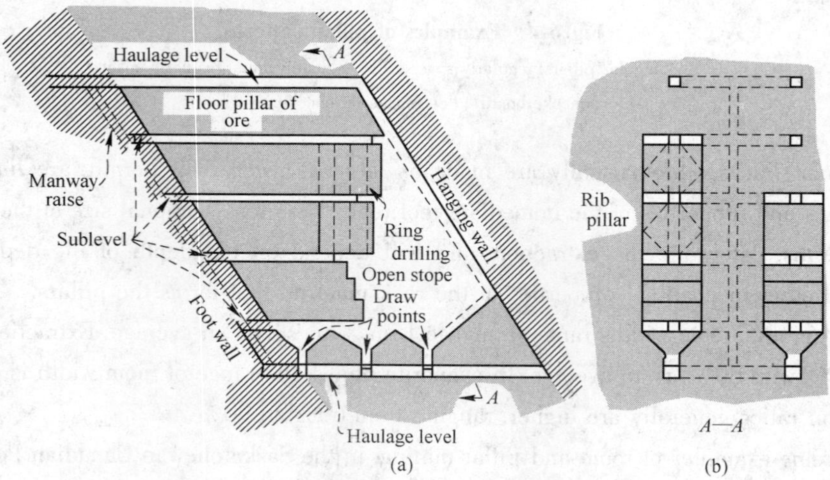

Fig. 6-6 Sublevel stoping with transverse stopes in wide veins

Longitudinal stopes are developed in comparatively narrow steeply dipping deposits. The stopes run parallel to the strike of the deposit and are of indefinite length. The width of the stope (or span) is limited by the thickness of the deposit. Either a random or a regular system of rib pillars may be left in stoped areas. Stopes up to 21m wide have been mined in this manner. Floor pillars normally are regular, forming the top of the worked-out stope and the bottom for the main haulageways or levels. If the deposit is steeply dipping, these floor pillars support lateral loads. This stope design permits an extraction up to approximately 75%. Some dilution from wall rock may occur, and some sloughing of the floor pillar is permitted, but more general caving must be avoided.

For wide deposits (greater than 21m) transverse stoping usually is employed. In this method, the stopes which run perpendicular to the strike of the deposit are limited in length to the thickness of the deposit. These stopes usually are outlined by a regular system of rib pillars, with floor pillars being the same as in longitudinal stopes. The spacing of the rib pillars is determined by the ability of the ore to form an unsupported span, but rarely is greater than 21m. If the deposit dips steeply, the floor pillars provide lateral support. The extraction with transverse stopes generally is less than with longitudinal stopes because a greater percentage of the ore is left in the form of rib pillars, but less dilution from the sidewalls is experienced.

Stopes have been mined with spans from 12 to 15m in massive sulfides with hanging and footwalls of rhyolite, brecciated tuffs and aggregates or conglomerates; from 14 to 27m in massive sulfide ore with greenstone hanging and footwall; and up to 30m in iron sulfides with hanging and footwalls of a highly metamorphosed schist, graywacke and slate.

4. Shrinkage Stoping

Shrinkage stoping (Fig. 6-7) is used to mine narrow or wide veins, and sometimes bedded deposits that are steeply dipping. This mining method basically is an overhand stoping system in which a portion of the broken ore accumulates until the stope is completed. The increase in bulk as the ore is broken requires that some 30% to 50% must be "shrunk" periodically through chutes or drawpoints to maintain a working floor for additional mining. In general, the vein material must be strong enough to stand unsupported across the width of the stope and, when broken as ore, should not pack to the degree that it cannot be drawn. In vertical to near-vertical deposits, both the hanging and footwall rock should be relatively competent competent to prevent failure and excessive dilution of ore.

During the period the stope is being mined, both the hanging and footwall rock are stabilized to some degree by the broken ore in the stope. When the stope is mined out and while the remaining broken ore is being drawn, some sloughing from the hanging wall or footwall may occur, but usually the void created by the stoping operation remains open after the draw is completed. The rib and floor pillars remaining, therefore, provide support for the

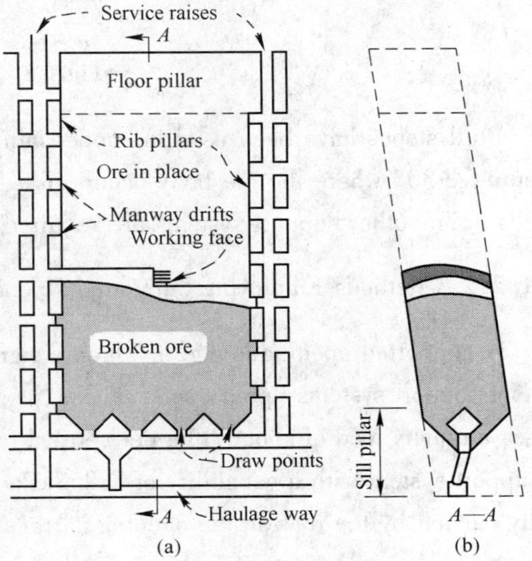

Fig. 6-7 Shrinkage stoping

hanging and footwall and can be in the same configurations as described in sublevel stoping.

Shrinkage stoping has been employed to depths greater than 762m. Extractions from 75% to 85% usually are obtainable. Spans of 10m have been mined in a fluorspar deposit with hanging and footwalls of limestone. Spans of 21m were obtained in veins of chalcopyrite and pyrite where the hanging wall was defined by a fault and the footwall consisted of an ore-grade cutoff within the veins. The rocks surrounding the veins are slate, schist and graywacke.

5. Stull Stoping

Stull stoping is a method which employs systematic or random timbering (stulls) placed between the foot and hanging walls of a vein. The vein may be flat-lying to steeply dipping tabular, or narrow in type and usually 3.7m or less in thickness (Fig. 6-8). The stulls provide the only artificial support and usually require that the hanging wall and sometimes both the hanging and footwall be moderately competent as, for example, thin-bedded or partially bonded, jointed and fractured rock types.

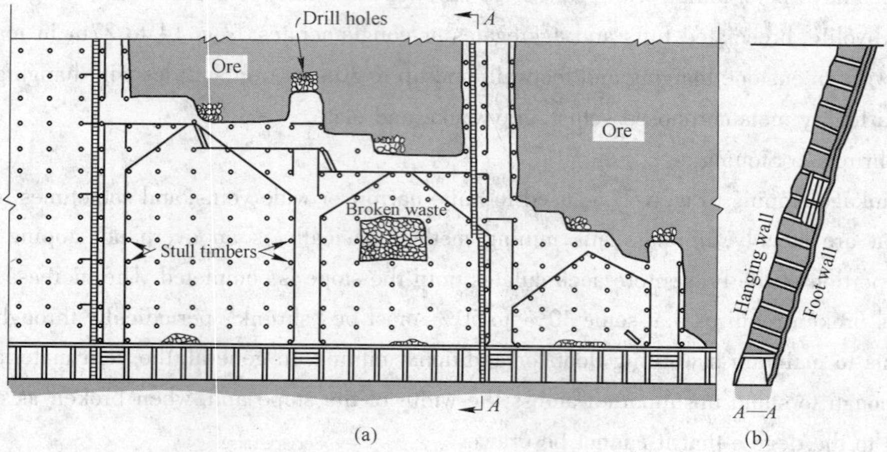

Fig. 6-8 Stull stoping

Stull stopes have been used in copper mining at a depth of 1,067m in a bed 3 to 3.7m thick dipping ±30° where the ore body occurs in a felsitic conglomerate. For mining thickness that exceeds 3.7m, other support systems are required.

6.1.2.2 Methods Employing Supported Openings

A supported opening is one in which a significant part of the incumbent load is carried on artificial support systems (props, sets, chocks, packs, backfill, etc.). Because the overlying cover imposes a gravity load of about 7kPa per vertical foot, an opening at a depth of, say, 152m will require a support system with a capability of 3,858kPa for total support if no part of the gravity load originally carried by the rock in the opening is transferred to the rock surrounding the opening after excavation. Support systems with this capability generally are impractical except in the later stages of mining with backfill. Even with maximum utilization, stulls, sets and other light support probably will not carry more than a few percent of the gravity load. Hydraulic props (jacks) can provide a greater support. For an opening at a depth of 152m, 163t hydraulic props placed on 1,067mm centers will support about 36% of the overburden weight. Chocks and pack walls, if placed close enough together, may support up to 50% of the gravity load in large stoping excavations. Backfill, on the other hand, can support 100% of the overburden weight after the surface subsides and compacts the fill material.

1. Cut-and-Fill Stoping

This mining method is best suited to vein or bedded deposits that dip at an angle greater than the angle of repose of the broken rock (Fig. 6-9). The ore may be massive, thick-bedded and partially bonded jointed material, but should be sufficiently competent to maintain stope-width spans. The surrounding rock, mainly the hanging wall, usually is of a type that will not stand for a long period without support.

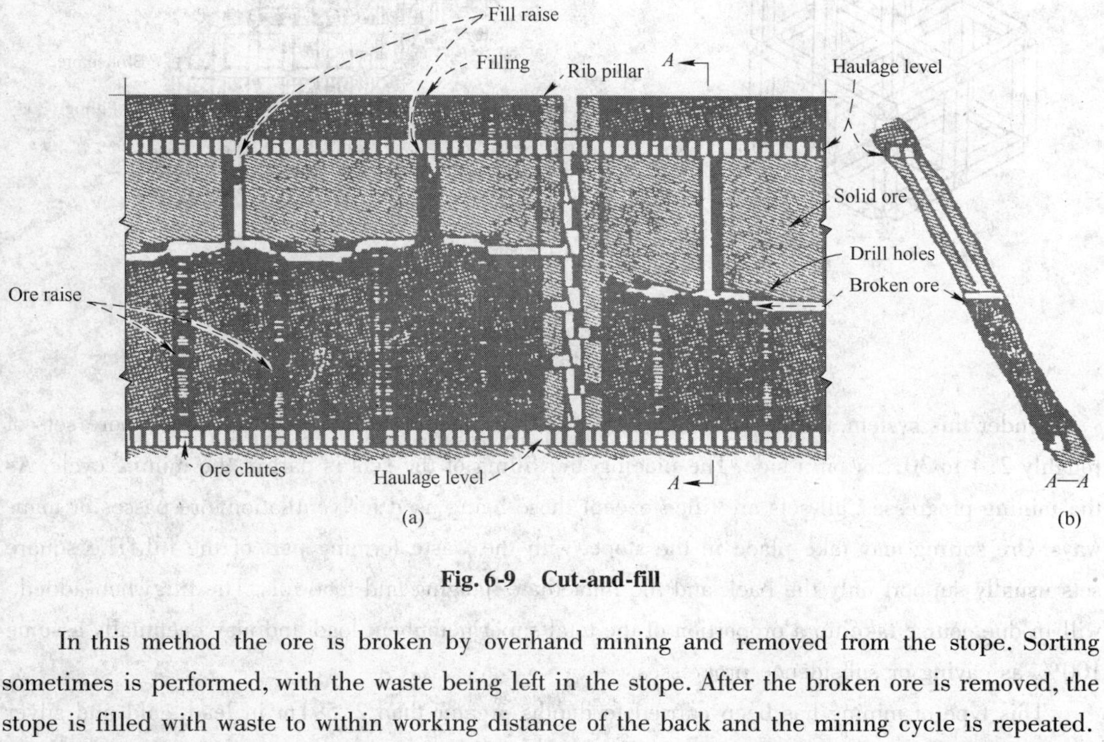

Fig. 6-9 Cut-and-fill

In this method the ore is broken by overhand mining and removed from the stope. Sorting sometimes is performed, with the waste being left in the stope. After the broken ore is removed, the stope is filled with waste to within working distance of the back and the mining cycle is repeated. The waste fill may be broken rock, sand and/or gravel, soil or classified mill tailings. Pillars, if any, usually can be recovered and extraction may be near 100%.

If the vein being mined is extremely narrow, a method of mining called "resuing" may be used. In this system, the vein and a waste wall are broken separately to provide a working width and the waste is left in the stope for filling.

Cut-and-fill stoping has been conducted at depths of 2,591m in quartz veins 3m wide, with the surrounding rock being a lava schist and porphyry. Also, this method has been used at depths of 1,097m in quartz and pyrite veins varying from 1.5 to 30m wide.

2. Square-Set-and-Fill Stoping

Although square-set-and-fill stoping (Fig. 6-10) can be employed to mine almost any type of deposit and in most rock types, it generally is used for deposits where the ore is structurally weak and where faulting and fracturing of the surrounding rock has resulted in it also being very weak. The method is adaptable to deposits with irregular boundaries and is extremely flexible where ore varies greatly in short distances. This method can be applied when all others have proved inadequate, and insures a recovery approaching 100%. However, it is high in cost of materials and labor.

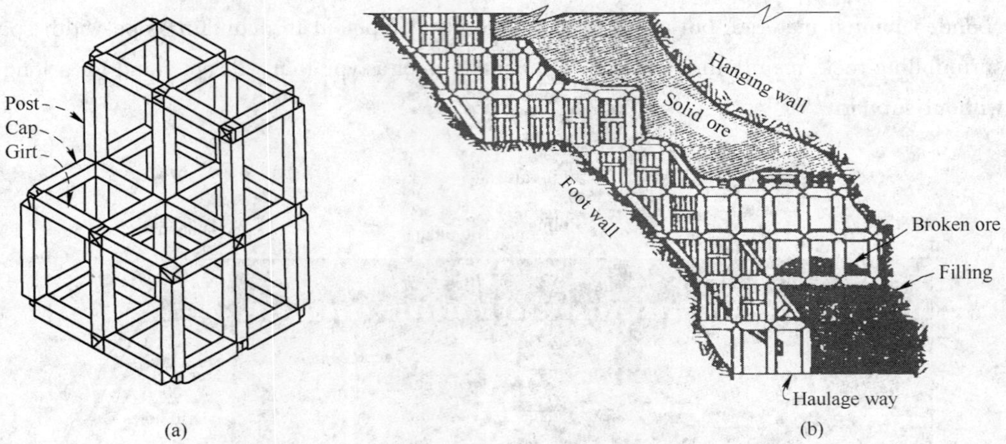

Fig. 6-10 Square-set stoping

(a) Square-set timbering; (b) Vertical transverse section

Under this system, the ore normally is mined from hanging to footwall for one or more sets of roughly 2.4 to 30.5m on a side. The placing and filling of the sets is part of the mining cycle. As the mining progresses, all sets are filled except those being used for ventilation, ore passes or manways. Ore sorting may take place in the stope, with the waste forming part of the fill. The square sets usually support only the back and the immediate hanging and footwalls. The fill, when added, will in due course take up a proportion of the total superincumbent load and may eventually assume 100% as caving or subsidence progresses.

This type of mining has been carried to depths greater than 2,591m in lead, gold and silver mines with hanging and footwalls of schists, porphyrys, highly altered quartzites, shales, limestones and granites.

An underhand-fill method of recent development, used with or without cemented fill, appears competitive with the square-set-and-fill method.

3. Longwall Mining

Longwall mining, in its original concept, is employed primarily to extract coal, although this procedure, with modifications, has been used to mine metallic minerals—for example, uranium and gold ore. The method is adaptable to deposits ranging in thickness from 0.9 to 2.4m, dipping at less than 12°, and lying at depths up to 914m or more, provided the rock materials overlying the deposit are generally thin-bedded, relatively incompetent and cave freely and completely behind the prop line (Fig. 6-11). Also, the rock in the floor should be sufficiently competent to support the prop loads.

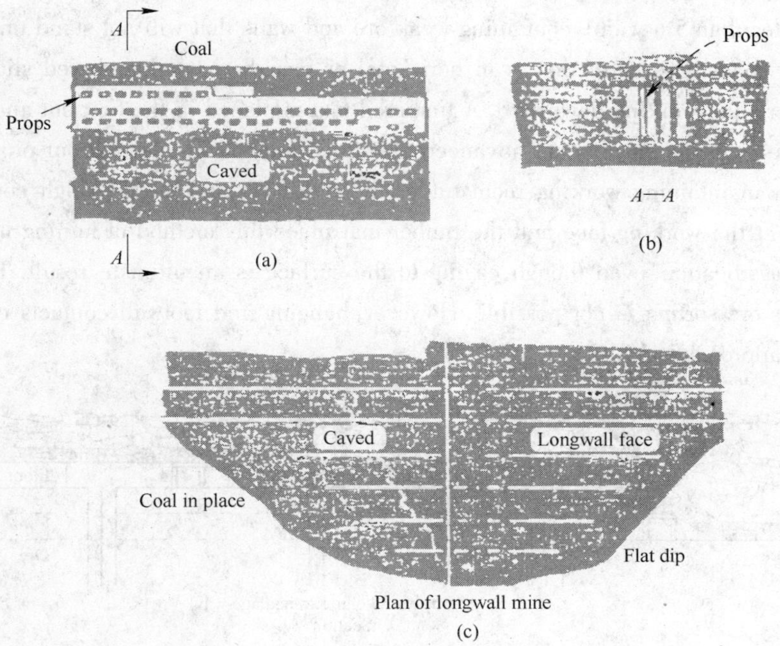

Fig. 6-11 Longwall mining

Because a massive system of props is used to support the roof over the face and working areas, longwalling is classified as a supported stoping method, although caving occurs in mined-out areas.

A sequential arrangement of stope faces on any dip to give a uniform pressure distribution, generally referred to as modified longwalling, is regarded as a method of ground control rather than a mining method.

In a longwall coal mine the face is fragmented with cutters, rippers, or plows, and the coal is transported from the face by a conveyor. Usually an extraction approaching 100% is possible. Because of virtually complete extraction, the overlying rock caves into the mined void, and the resulting surface subsidence is relatively uniform and complete. However, in some European coal mines in which longwalling is employed, caving and hence surface subsidence is limited by placing rock packs in mined areas.

4. Shortwall Mining

A shortwall method is utilized in the same type of deposit and in the same rock material as long wall mining. The primary difference in the two methods is the length of the working face. In shortwall mining the maximum working face is normally 46 to 49m. Safety and mining laws in the

United States require this shortening of the mining face. Also, the shorter face is more flexible, particularly in areas of bad roof. The equipment utilized and the results of this method of mining are the same as given for longwall mining.

5. Top Slicing

Top slicing is essentially a method for mining massive, thick-bedded or wide-veined deposits (normally greater than 5m wide) containing weak ore and walls that will not stand unsupported except over short spans. Ore extraction is in horizontal or near horizontal timbered slices starting at the top of the ore and working downward. A timbered mat is placed in the first cut and the overburden is caved. As subsequent cuts are advanced caving is induced by blasting out props behind the face, but always maintaining working room under the mat (Fig. 6-12). The high concentration of timber support at the working face and the timber mat place this method of mining under the supported stope classification, even though caving to the surface is an ultimate result. The method is not selective, as ore sorting is not possible. However, hanging and footwall contacts can be irregular. Extraction approaches 100%.

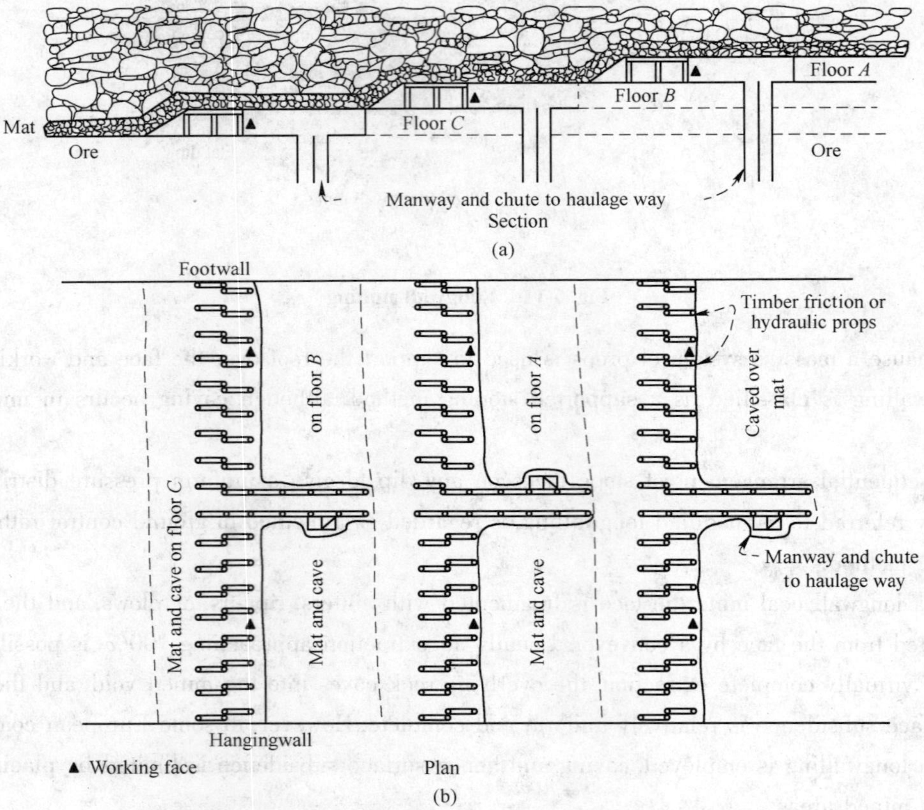

Fig. 6-12 Top slicing with mat

Top slicing has been used successfully at depths of 762m in a deposit of soft hydrated hematite with a friable and highly fractured cap of chert jasper overlain by glacial fill. The system has worked equally well in other iron ores at depths of from 46 to 122m.

6.1.2.3 Caving Methods of Mining

Three caving methods are generally recognized: top slicing, sublevel caving and block caving. In all cases, the general requirements for use of the method are massive-type mineral deposits of large horizontal area, such as thick beds, masses or wide veins. The ore should be weak, or if hard it should be thoroughly fractured with weak bonding, or a combination of these factors. Overburden may range from firm rock to glacial drift but must cave and follow the ore down as the ore is removed. Top slicing, because of timber requirements and method of support, has been described under supported stopes.

The rock materials in the ore body and overburden should be competent but highly jointed or fractured, and with virtually no bond across joint planes. These rock materials, due to the fracture pattern, should cave when support is removed from a sufficient area, which may range from 3m to more than a 61m square. In some rock types, assistance is needed to initiate the cave, such as outlining the area to be caved by longhole slots and presplitting. After the area has been detached from the surroundings, the rock should cave under its own weight, move by gravity and form fragments that are small enough to be handled conveniently. The method is most applicable to large low-grade ore deposits. Since the cave progresses to the surface, the method can only be employed where such disturbance can be tolerated.

1. Sublevel Caving

Sublevel caving can be used to mine massive or large pockets of ore and thick-tabular or wide-vein deposits that dip steeply. The method usually is employed to mine ore bodies of large horizontal extent, often below an open pit where irregularity of the walls or other factors make it preferable to block caving. The rock material in the deposit should be moderately competent, such as a jointed or fractured rock with some joint strength. The rock should not be free-caving, but when it is broken small fragments should be formed. This method, within limits, also can be utilized to mine soft sticky ores which have a tendency to repack. The rock material in the capping should be jointed and a tendency to repack. The rock material in the capping should be jointed and fractured, with partial bonding, and follow the ore cave without undue dilution or delay. When the overburden fails, small to medium-sized fragments should be formed but with very little fines.

The sublevels are driven between and parallel to the main levels, with the distance between the sublevels varying between 6 and 12m, giving a 6 to 9m column of rock to be fragmented by blasting (Fig. 6-13). As the rock is fragmented

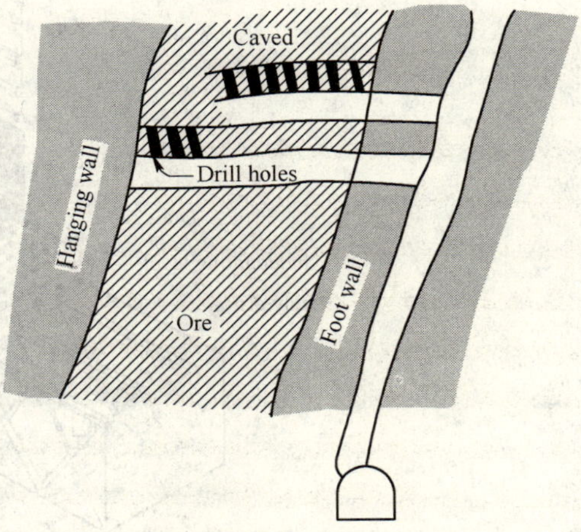

Fig. 6-13 Sublevel caving

and collapses into the sublevel for removal, the overburden immediately caves onto the fragmented rock. This method requires the mining of the mineral deposit from the top down. Sublevel caving is nonselective, permitting no sorting in the stope. The hanging wall and footwall can be irregular. This method generally is used to mine low-grade deposits. A 90% recovery with substantial dilution is considered normal.

Sublevel caving has been carried out in ore deposits which contain soft hematite where the footwall is quartzite and siliceous slate, and the hanging wall is a cherty iron formation which is fractured with partial bonding.

2. Block Caving

Block caving is normally conducted in massive and disseminated low-grade mineral deposits of large horizontal dimensions which are structurally weak. The rock material in the deposit and overburden should be incompetent and should thus cave freely. Highly jointed, or fractured or thin-bedded rocks with very low bond strength across joints, fractures and bedding planes are typical caving material. The type of rock material in the overburden normally is not as important as the joint or fracture spacing, the degree of alteration in the rock and the lack of bonding in fractures. These rock materials should break but not repack. This allows for the grinding action of the ore during the caving cycle.

Block caving is a low-cost high-production method. Main haulageways and drawpoints are constructed below the block to be caved. The height of the block normally is greater than 30m (Fig. 6-14). The block is undercut and permitted to collapse, with the fragments being drawn off periodically. The initial block to be caved may require some loosening from the surrounding rock to initi-

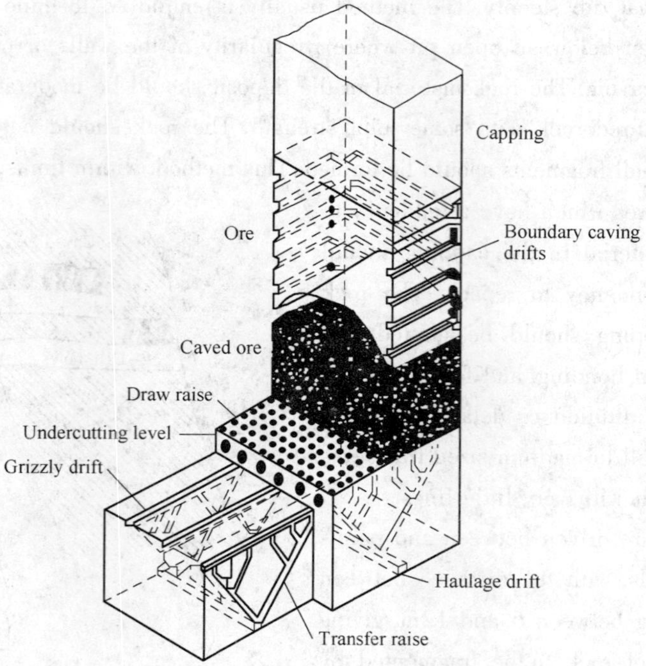

Fig. 6-14 Block caving

ate the caving action. Extraction of the mineral deposit with this type of mining is approximately 100%, although dilution may require an earlier cutoff and reduce recovery.

This type of mining is utilized in the porphyry copper deposits in the South-western United States, the molybdenum deposits in Colorado, the iron deposits in Northern Michigan, and in asbestos deposits in Canada. It has been carried on successfully at depths greater than 610m.

6.1.3 Selection of Surface Mining Methods

A surface mine is an open-air excavation for removing minerals. Surface mining can be employed to extract metallic or nonmetallic mineral from any near-surface deposit in any rock type. Near-surface generally implies depths less than 152m. The size of the deposit may range form a few tonnes (for example, small pockets of barite mined in eastern Missouri) to more than 102 million metric tons (for example, the Bingham Canyon copper mines in Utah). The larger deposits usually are of considerable areal extent. The surface mining methods listed in the following tabulation fall into one of the three basic types: placer, open pit and glory hole, or combinations of these.

(1) Placer Mining: 1) Panning and sluicing; 2) Hydraulicking; 3) Dredging.

(2) Open Pit: 1) Single-bench; 2) Multiple-bench; 3) Strip mining; 4) Quarry mining.

(3) Glory Hole.

Surface mining permits a wide flexibility in production, which includes the ability to mine selectively and the potential for 100% extraction of ore within the pit limits. Fewer men are required, since mechanization permits high unit production and greater safety than underground mining.

Problems in surface mining include adverse weather in some locations, and undesirable environmental problems, such as surface scarring, dust, noise and blasting vibrations, as well as waste disposal.

6.1.3.1 Placer

Placer mining is effected by the concentration of minerals from detrital materials by selective settling in running water. This mining operation usually requires that the ore deposit be in or near water and on or near the ground surface. The preceding tabulation lists three common methods of placer mining—panning and sluicing, hydraulicking and dredging.

1. Panning and Sluicing

Panning was used in the early days of gold mining in the United States in placer deposits where water was plentiful and the ore (gold, silver, precious stones, etc.) was concentrated in layers or pockets. Panning can be utilized only when the ore is heavier than the gangue (waste rock) and when high production is not required. Being a low-production method, panning is used today primarily for exploration purposes in tracing placer deposits or ore to the vein or other source.

Sluicing, used in the early days of gold mining, has been largely replaced by higher-production methods. It is utilized today mainly for exploration. The placer deposit for this type of mining can be 1.8 to 2.4m thick and also may be dry. However, water and a trough-shaped box (sluice box) are needed for separating the ore and waste, and a slope is needed to carry away the water and waste. The size of the equipment and the angle of the slope depend on the size of the deposit, the time needed for the gravel to disintegrate, the flow of water available and the maximum gravel or boulder size expected.

Because most placer deposits in the United States suitable for panning or sluicing have been exhausted, environmental problems usually associated with this mining method are no longer of concern.

2. Hydraulic Mining

Hydraulic mining is employed in larger placer deposits that generally contain gravel and boulders. Large quantities of water under pressure directed through pipes and nozzles (giants) are utilized to disintegrate the deposit. The water should have a head high enough to cave the gravel and move it to sluice boxes. Additional water, referred to as bank or by-wash water, may be needed to operate these boxes.

A face height of 5 to 15m is normal. The maximum face should be no higher than 45 to 61m. For best results, the bedrock should slope at least 2% for fine gravel, 4% to 5% for medium gravel and greater than 5% for coarse material. Production with this mining method is limited by available water, thickness of the deposit, boulder size and bedrock slope.

Environmental problems with this mining method are water runoff and tailings disposal. Also, since hydraulic mining eventually disturbs a large surface area, land restoration becomes a problem.

3. Dredging

Dredging is the underwater excavation of a placer deposit of detritus-type rock material. For this mining method, the deposit usually is low-grade and large in both areal extent and thickness. The deposit may be an active or extinct stream or river bed. The best surface slope of the deposit is 2%, with a maximum of 6%. If the bedrock is flat and hard so no mineral, etc., can sink into it, or soft and fractured so it can be dredged, recovery will be increased.

The dredge is basically a continuous large-volume digging machine attached to a gravity concentrating facility (jigs, etc.) with waste removal systems, all contained on a floating platform.

Dredges can remove material from 4 to 30m below the surface of the water and have been used successfully to a depth of 49m in Malaysia. Bank heights (Fig. 6-15) above water level normally should be no more than 4.6 to 7.6m, but may go as high as 12 to 15m for the large dredges. If the

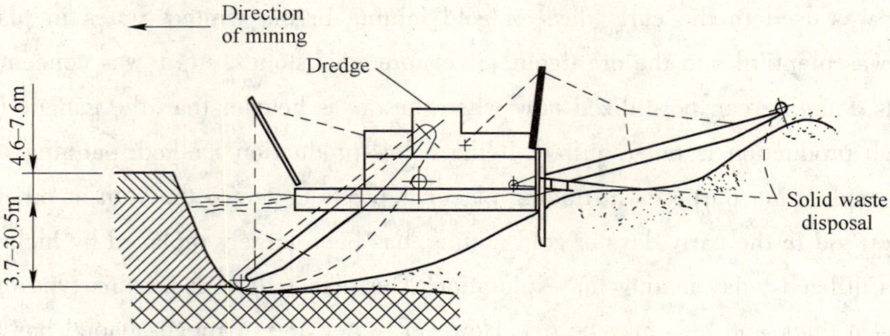

Fig. 6-15 Dredging

bank is much higher it should be stripped in some manner to avoid cave-ins. The clay content in the gravel, the boulder size and the amount of water available to float the dredge determine the tonnage or size of the dredge.

Problems in dredging are containing sufficient water in a pond to float the dredge and having enough clear water to beneficiate the gravel. Owing to the large size of the operation in dredging, tailings disposal and restoration of the land and water are of major concern.

6. 1. 3. 2 Open Pit

Open pit mining, sometimes referred to as open cut mining can be employed in any mineral deposit in any rock type lying near or on the surface where the excavation is open to the sky and weather. These methods are best suited to ore bodies of substantial horizontal dimensions which permit high rates of production and thus a low cost. Although stripping and quarrying are forms of open pit mining, "strip mining" usually refers to the mining of coal and "quarry mining" is related to the production of nonmetallic materials, such as dimension stone, rock aggregates, etc.

Factors that determine the pit layout include orientation of the jointing and other fractures in the deposit, the stripping ratio, the weather conditions, rate of production required and the equipment available. Also, at some depth and for every type of deposit, there is the option of mining by either surface or underground methods. Usually, economic factors dictate this choice. Some mines have employed both surface and underground methods. For example, the Crestmore mine, Riverside, Calif. , changed from surface to block caving, and then to room-and-pillar mining as the depth of the operation increased. The Morenci mine, Morenci, Az. , and the Jennifer mine, Boron, Calif. , changed from underground to open pit mining. The relative improvement of surface excavation equipment over underground mining equipment has been a factor that favors open pit and strip mining methods.

Four types of open pit mining are as follows: (1) single-bench, (2) multiple-bench, (3) strip and (4) quarry.

1. Single-Bench

In an open pit mine a bench forms a single level of operation above which minerals and waste materials are excavated from the bench face (Fig. 6-16). Single-bench open-pit mining can be em-

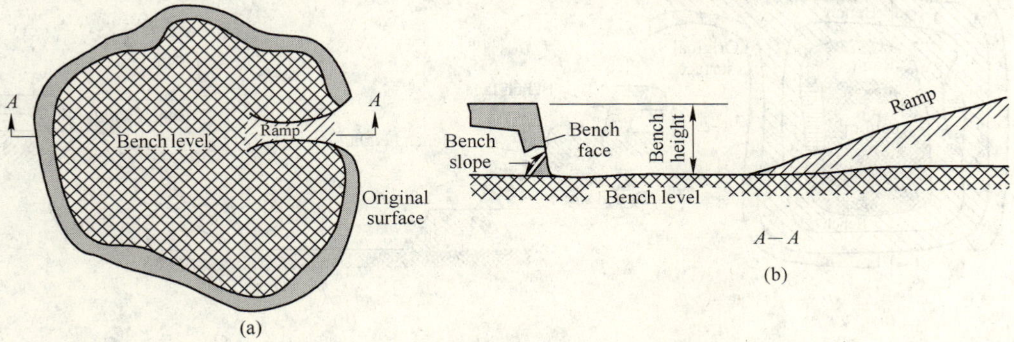

Fig. 6-16 Single-bench open-pit mine

ployed to mine any comparatively shallow mineral deposit in any rock type. Thus, quarry and strip mining may be single-bench operations.

The maximum stable bench height and bench slope depend on the rock materials that form the bench (may be specified by law in some cases). Bench heights normally range from 4 to 8m for weak to moderately strong rocks, but may approach 61m in very special cases. There is no physical limit on the bench area.

Typical deposits mined by this method are sand and gravel beds, coal seams with limited overburden (see strip mining) and near-surface exposures of dimension and aggregate stone (see quarry mining).

Production by this method is limited only by the capacity of the equipment that can be employed in the pits and by the number of areas along the face that can be excavated simultaneously.

Waste disposal, noise, dust, blasting vibration and surface restoration create environmental problems.

2. Multiple-Bench

Multiple benches normally can be employed in any massive, thick-bedded, wide-vein or tabular deposit that lies at or extends to a depth greater than that suitable for single bench mining. The rock material may be of any type sufficiently strong to permit the development of benches of economic height, and may consist of semiconsolidated (playa) to weathered, altered and often highly fractured types, that is, rock material requiring only minimum blasting to permit excavation.

If the pit is over 8 to 15m deep, more than one bench probably will be needed. With more than one bench, the bench width (Fig. 6-17) may vary from 6 to 23m depending on the size of the haulage equipment and on the rock material. Benches normally are used for roadways, either forming a spiral to the bottom of the pit or with ramps between horizontal benches. Bench widths or berms also are designed to provide protection for men and equipment from small slope failures (slides). Bench slopes (Fig. 6-17) usually are steeper than the pit slope, because rock can maintain a more nearly vertical wall for short heights. Pit slopes may vary from 20° to 70°. During the final mining cycle before abandonment, the pit slope sometimes is steepened to increase recovery.

Problems of an environmental nature arising from multiple-bench mining are waste disposal,

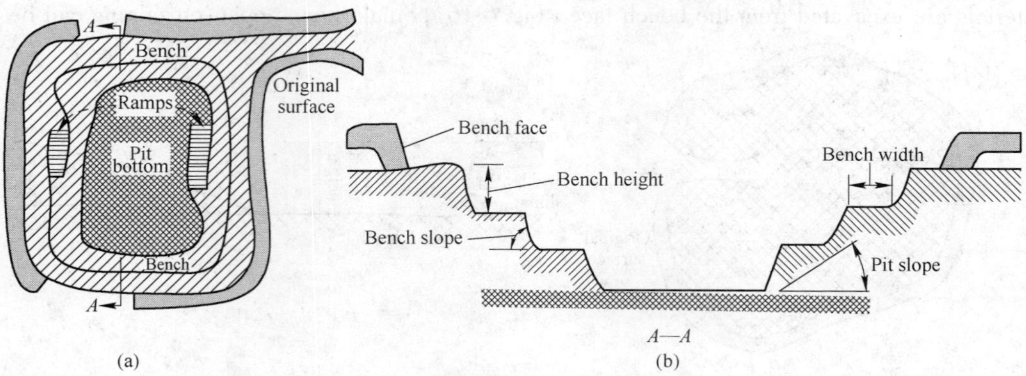

Fig. 6-17 Multiple-bench open-pit mine

noise, dust, blasting vibrations and land restoration. Water may or may not cause environmental problems.

There has been adverse reaction to open pit mines because of the disturbance to the environment. However, most high-production underground mines, such as block caving operations, cave to surface almost immediately and produce scars or otherwise make the surface unusable. In many instances, land restoration and utilization are easier and less costly to effect over surface excavations than over underground mined areas that eventually subside.

3. Strip Mining

Strip mining is the term mainly applied to the mining of near-surface seams of coal. However, other mineral deposits which have low cohesive strengths also can be mined by this method. Most stripping is effected in bedded sedimentary formations. Blasting may or may not be required, depending upon the type of overburden.

Coal seams mined by stripping range from approximately 0.6 to 9.1m or more in thickness. Usually, thicker and multiple seams are mined by benching. With modern equipment, overburden to coal ratios as high as 30 : 1 and depths ranging up to 46m have been stripped where overburden conditions are favorable. The high stripping ratios also depend on mining costs and equipment efficiencies, which vary widely and in a majority of cases control the relationship of minable coal and overburden thicknesses.

Strip mining usually is accomplished by removing the overburden and coal from a strip across one dimension of the deposit. A parallel strip is then excavated in the opposite direction and the overburden or waste rock is placed into the strip previously mined. This cycle is repeated as often as the deposit allows (Fig. 6-18).

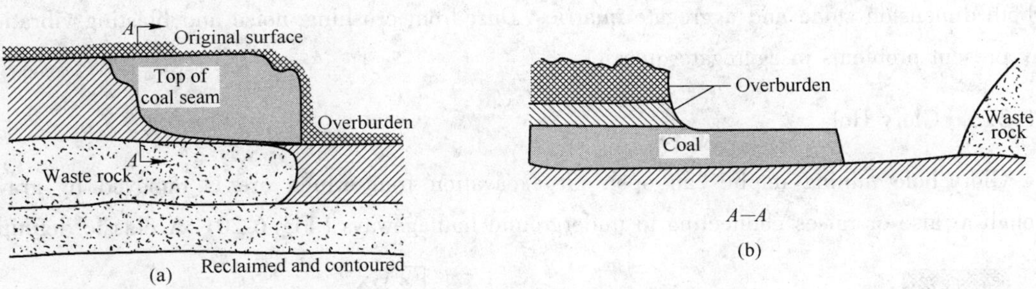

Fig. 6-18 Strip mine

Stripping equipment is normally quite large and may either discharge the waste material directly into the previously mined strip or use a conveyor system. Examples are wheel excavators ranging from less than 0.8 to greater than $4m^3$ per bucket; dragline buckets ranging from approximately 5 to over $168m^3$. Shovel buckets may be as large as $153m^3$. Mining the coal is accomplished with shovel buckets normally rated from 5 to $9m^3$.

In strip mining the maintenance of a sidewall is not as critical as in open pit bench mining. However, high waste piles do present slope failure problems.

The environment (water, temperature range, weather conditions) , variation in rock properties, stripping ratio and transportation are the main problems encountered in strip mining. Final waste disposal and land restoration can be included in the mining cycle, but at an increased cost.

4. Quarry Mining

Quarrying is the used to describe the surface mining of rock, such as marble, granite, limestone, slate, etc. , that are valuable for either their mechanical or chemical properties. In this type of mining the deposit usually is either massive, bedded or lenticular, and is suitable for bench mining. Most quarries are in sedimentary rock (limestone) . However, some metamorphic (marble) and igneous (granite) rocks also may be mined in this manner.

There are two basic types of quarries: dimension stone and broken stone (i. e. , aggregate and chemical limestone) .

Dimension stone quarries normally have benches with vertical faces and the overall pit slope is steep (Fig. 6-19) . The rock therefore must have bonding across any fracture or joint planes producing a relatively high cohesive strength. The stone generally is broken loose by some manner of cutting instead of blasting. This is done to preserve its shape and strength. Bench heights can range as high as 61m. Production normally is very selective and in limited quantities.

Quarry mining of aggregate or chemical stone usually is done by blasting to fragment the rock. The degree of fragmenting depends on the product desired. The rock therefore can have a low cohesive strength. Production is higher and less selective than for dimension stone quarries. Most stone quarries require some overburden removal.

Water seepage, land restoration and overburden disposal are environmental problems common to both dimension stone and aggregate quarries. Dust from crushing, noise and blasting vibrations may present problems in aggregate quarries.

6. 1. 3. 3 Glory Hole

Glory hole mining implies an open pit excavation from which ore is removed by gravity through a raise or raises connecting to underground haulageways (Fig. 6-20) . A classic definition

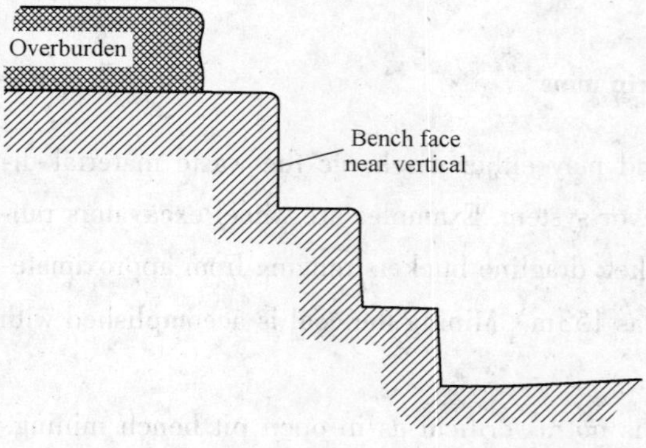

Fig. 6-19 Dimension-stone quarry

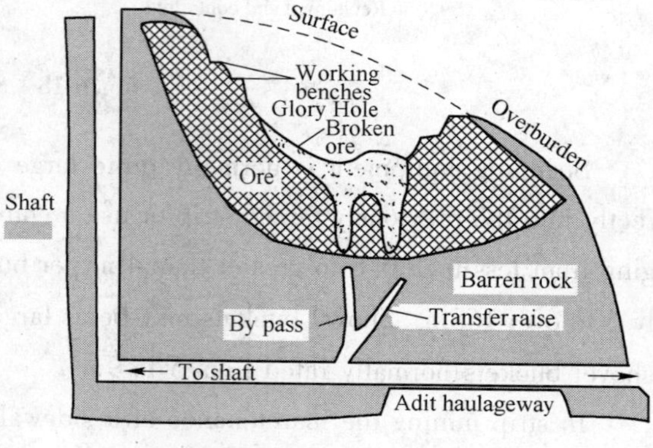

Fig. 6-20 Glory hole mining

usually describes an operation or mining method where ore around each raise is excavated so that it falls into a raise by gravity, resulting in a funnel-shaped configuration which may also be called a "mill-hole" and the overall process "milling". Modern open pits may haul to ore passes connected to skip loading facilities at or below pit limits. However, this or similar modifications do not qualify as true glory hole mining.

A glory hole method may be used to mine almost any type of deposit that can be mined by open pit methods and some narrow steeply dipping veins can utilize this method. The rock material in the deposit can be of any type but should not have a tendency to pack in the drawpoints. In this type of deposit the pit may become deep, narrow and long (Fig. 6-20). The sidewalls may or may not be benched and therefore should be massive or, if fractured, of high cohesive strength. Large open pits which are reaching their lower boundaries and are starting to mine underground might utilize this approach as a changeover or interim method.

Problems arising in this type of mining are water causing the ore to pack in the draws, and sidewalls sloughing in. Depending on the size of the operation land restoration, tailings disposal, dust, noise and blasting vibrations could be environmental problems.

6.2 Underground Mining

6.2.1 Shrinkage Stoping

Shrinkage stopes are overhand stopes in which broken ore is used for a working platform and to support the walls of the stope. As broken rock requires a larger volume than solid, some muck must be drawn off as the stope advances. Hence the name "shrinkage".

Shrinkage stopes are most used in steeply dipping vein-type deposits where the ore is strong enough to stand with no support and the walls with very little. Slabbing and minor weakness in the country rock can be tolerated as long as resulting dilution is not a serious problem, but severe slabbing will plug the drawpoints and squeezing will bind the broken ore in place. Ore must be strong because it is uneconomical to support the back. Forty to 50° is the absolute minimum dip and when this is flat, ore draws from the hanging wall leaving a poor working surface—if it draws at all. For efficient mining the dip should be steeper than 60°. In the past, some massive deposits have been mined with panels of vertical shrinkage stopes. The intervening pillars then were mined by any of a variety of supported methods. Current economics make this unattractive.

Due to rising costs and a shortage of skilled miners, shrinkage stoping is being replaced by higher production methods wherever possible in the United States and, according to M. F. Goudge, in Canada also. Normally, the method now is used only in relatively high-value narrow deposits which cannot be mechanized, or in small operations where the expense of mechanization is too great.

(1) Levels——The variety of shrinkage-stoping configurations reflects both the necessity of combining exploration and development work in a vein-type deposit, and the desire to protect necessary development workings while minimizing unrecoverable pillars of ore. Levels may be driven at

vertical intervals of 30 to 152m. The interval should be as large as the ore warrants, but maximum stope height is determined by draw control and stopes over 61 to 91m high tend to draw erratically. If two stopes are developed above each other between a pair of levels, then a reasonable maximum level interval will be 152 to 183m. To the extent that one is confident of ore consistency, levels may be driven in the footwall, provided the vein is straight enough to permit a reasonably constant pillar between vein and drift, based upon core-drilling information.

(2) Raises——Vertical access into the stope may be through cribs, single stull-laced raises, or conventional raises driven in the vein or footwall. Stopes with a short vertical rise may be developed by cribbed raises through the broken muck. After 3 to 5m, the cribbing must be bolted to the footwall with a crib hanger because the movement of shrinkage muck will endanger the raise (Fig. 6-21). A more-secure raise may be constructed with stulls and lagging at the ends of the stope. The raise may be divided into two compartments, one containing service lines and a ladderway, and the other with a small skip (Fig. 6-22). For larger stopes, a timbered raise between pillars of ore is required. Dogholes a round or two deep should be cut into the vein on each rib at regular intervals for access to the stope as the stope advances (Fig. 6-24). The raise may be driven in the footwall of the vein, in which case the dogholes would be cut from the hanging wall through the vein (Fig. 6-25).

6.2.1.1 Stope Preparation

The primary purpose of the stope preparation phase is to construct a means of drawing off the shrinkage muck. Stope preparation normally begins after ore has been blocked with levels and raises, although this is not necessary.

The simplest method of developing a shrinkage stope is to take one or two cuts out of the back of the drift, install timber and chutes, and provide access through cribbed raises at the ends of the stope (Fig. 6-21). Some of the disadvantages of this method are large timber requirements, high cost of level maintenance, and danger to haulage and traffic from slabs falling from the stope after the muck has been drawn. Furthermore, cars must be present when miners wish to draw off excess muck, thereby inhibiting orderly haulage. Finally, oversize muck must be bombed in the chute, which may damage both the chute and nearby timber.

An improvement on the previously described method is to construct chute cutouts and chutes on about 7.6m centers and hog over between the cutouts to start the stope (Fig. 6-22). With a substantial chute, there is considerable support over the haulage level. To elimi-

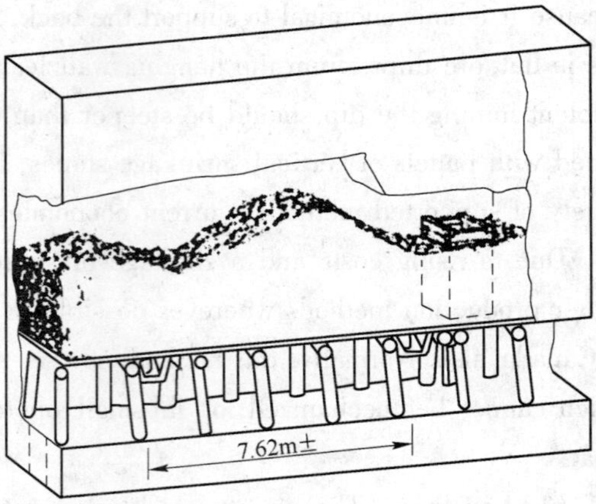

Fig. 6-21 Cribbed service raise and back-timbered shrink stope

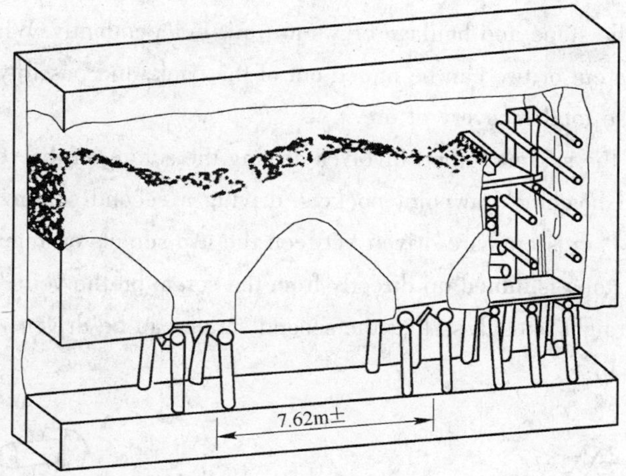

Fig. 6-22 Laced service raise and hogged-pillar shrink stope

nate extensive timber and chute-framing costs, some operations have tried a form of drawpoint loading. Pockets are cut into the walls on regular centers and hogged back onto the vein and together. The pockets are deep enough for a small mucking machine to pick up broken ore and load it back into waiting cars. Although the drift is protected by a pillar of ore and there are no timber costs, there still are disadvantages. Mucking machines must be moved, and empty cars block orderly haulage. Furthermore, when the stope has been drawn, slabs and loose material may roll out onto the haulageway. This system of development has been most used for panel shrink stoping in larger ore bodies where gathering drifts can be driven off the main haulageway and a mucking machine can serve several drawpoints off the gathering drift without being moved a great distance (Fig. 6-23). The most satisfactory system for a large mine makes use of a scram drift driven form the raise on the vein so as to leave a pillar over the haulage level. Drawpoints on regular centers are cut in the footwall of the scram and broken ore is slushed to a pocket and chute at the raise (Fig. 6-24). The

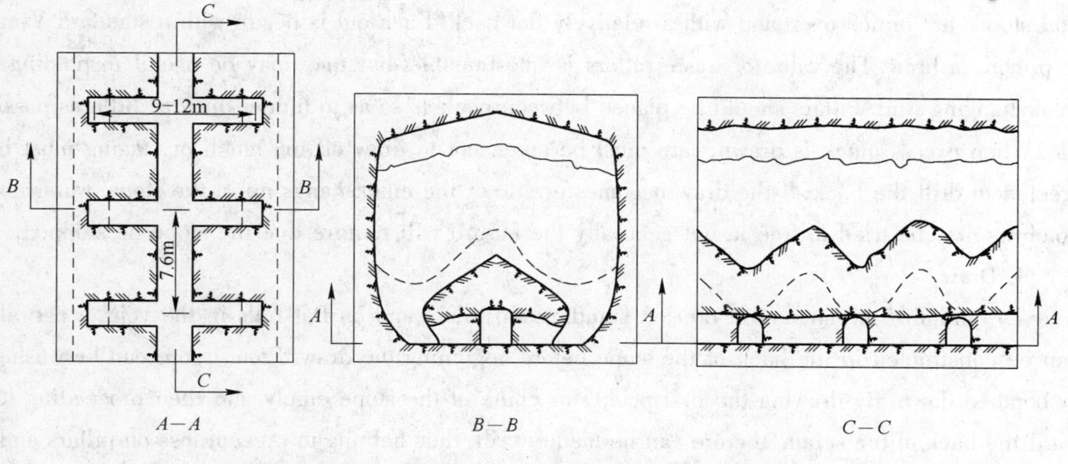

Fig. 6-23 Drawpoints on haulage level in a large ore body

resulting configuration will be slightly cheaper than constructing numerous chutes on the haulage level, and will permit the stope and haulage crews to work independently. When the stope has been mined out and drawn, a cut or two can be mined out of the back, and possibly one out of the bottom of the scram, to increase total recovery of ore.

A modification of the scram method involves driving the scram on the vein as above but then, rather than cutting and hogging drawpoint pockets, driving a second scram parallel to the first in the footwall (Fig. 6-25). crosscuts are driven between the two scrams on regular centers for draw of broken muck, and the stope is mined up directly from the scram on the vein. This is satisfactory only where the vein is straight enough so that the second scram can be driven with a constant 3m pillar between it and the first.

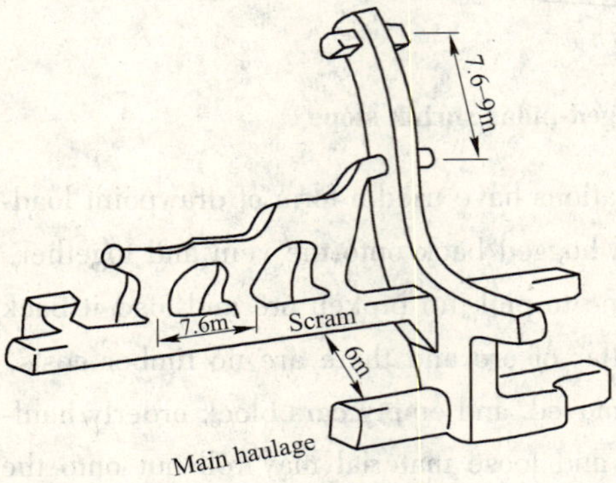

Fig. 6-24 Vein service raise and single scram with hogged pockets

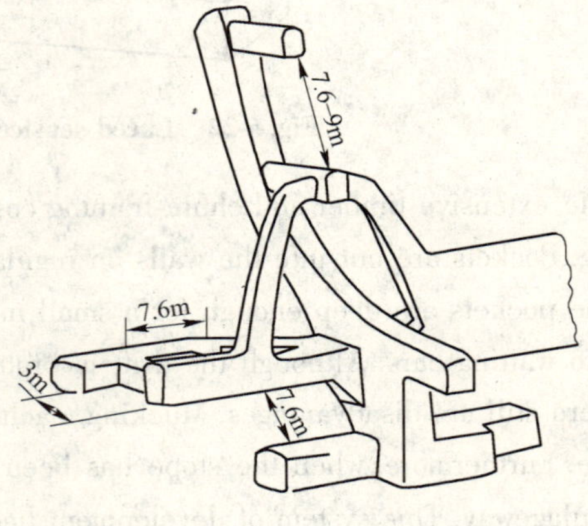

Fig. 6-25 Footwall service raise and double scrams with crosscuts on 7.6m centers

6.2.1.2 Stoping

Once the stope has been prepared, stoping is relatively straightforward. Almost without exception stopes are mined overhand with a relatively flat back. Each cut is begun with a standard V-cut to provide a brow. The value of waste pillars is questionable, but they may be useful in holding a weak hanging wall. Pillars should be placed between pockets so as to hinder draw as little as possible. When excess muck is drawn, care must be taken not to draw off too much or staging must be erected to drill the back. If the draw becomes erratic or the muck hangs up in the stope, water and bombing may be tried to free it, but generally the hazard will require that the stope be stopped.

1. Draw

To minimize the amount of ore left standing on pillars and on flat rolls in the vein, a catwalk can be constructed in the back of the stope before beginning the draw. From it, ore can be washed or bombed down. By drawing the first pocket or chute of the stope empty and then proceeding toward the back of the scram, the ore can be made to rill, thus helping to prevent loss on pillars and, in a scram stope, minimizing the slushing distance.

2. Broken-Ore Reserve

When a stope is near completion, it still will contain approximately 60% of the ore. It may represent hundreds of man-days of work, and for several months is a nonincome-producing investment. It must be considered a cost of operation. A system of accounts and projections detailing broken ore tied up in mining and available for draw must be maintained to account for this expense and provide an orderly feed to the mill. If permitted to stand in a stope for a long period of time, oxidation of the ore minerals may occur, making mill recovery difficult.

6.2.1.3 An Example of Shrinkage Stope

Idarado mine is in the southern San Juan Mountains of Colorado, between Red Mountain Pass and Telluride. It produces (1971) about 1,626t/d entirely from shrinkage stopes. The Montana-Argentine-Black Bear veins are being mined over a strike length of 5,486m and a dip of 914m. Dip of the veins varies between 45° and 90°, with most workings averaging 70° to 85°. Vein width generally is 1.5 to 2.1m, occasionally as narrow as 0.6 m and as wide as 7.6m. Replacement ore in the Telluride conglomerate has been mined to widths of 18m with no alteration of the basic stoping plan. Fig. 6-26 shows the generalized mine layout.

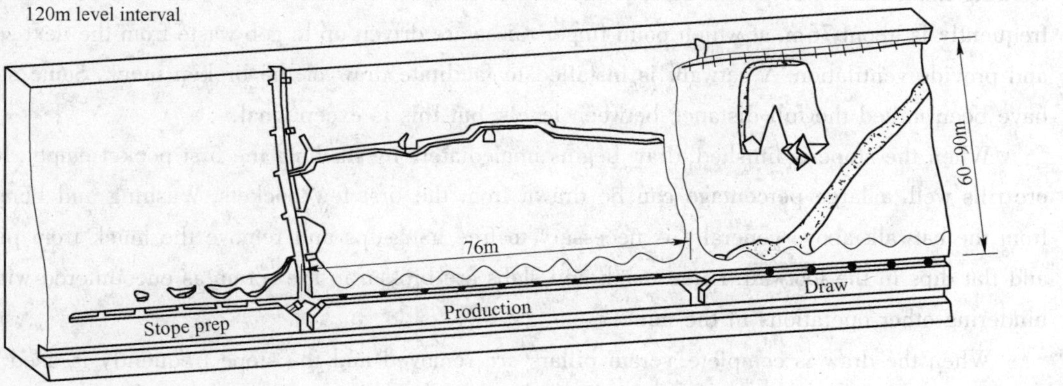

Fig. 6-26 General layout of shrinkage stopes at the Idarado mine

Levels are driven on the vein at 122 to 152m intervals. Footwall drifts have been driven, but with poor results due to variability of the strike and insufficient geological information from diamond drill holes. Raises are driven in the footwall of the vein with an Alimak raise climber, leaving a 3m pillar of country rock between the raise and the vein. Dogholes are driven from the hanging wall of the raise through the vein on 9m centers as the raise is being driven. Boring and reaming raises has been attempted but a tendency of the hole to steepen in dip makes the resulting raise difficult to use. Raises are driven on 122 to 152m centers.

Stope preparation begins with cutting a chute and pocket cutout, and constructing a slusher platform and grizzlies at the elevation of the scram sill—about 9m above the sill of the haulage drift. Scrams then are driven on the vein 61 to 76m in either direction. A scram is connected to a scram which has been driven toward it from the adjacent raise. At the breakthrough, a second chute and slusher platform are constructed to lessen the slushing distance. Pockets then are cut in the footwall on 7.6m centers, the first round being 1.8m high and 3.6m wide, and the second 2.4m and 2.1m. Three rounds then are blasted up from the back of the pocket to the vein, each successively wider. Care must be taken to air the pocket out after each hog round has been blasted. The fourth or fifth hog round should be on the vein and wide enough to break into the next pocket. The back then is blasted down to fill the pocket and, after the pockets are all hogged together, a leveling-up cut is taken out of the back. A finger raise is driven from the end of the scram up into the stope and a bearing set for the cribbed raise is constructed. At the font end of the stope, a finger raise is driven up into the first doghole of the service raise, and stope preparation is complete.

Stoping begins by taking a V-cut out of the middle of the stope. This provides a brow to which successive cuts are broken. Twenty to 25 holes are blasted at a time, but the last three rows of holes in each blast are not shot to provide holes through the shattered brow for the next blast. If a pillar must be left in the stope, it is placed between pockets, mined two cuts high and hogged over. Before each blast the pocket below should be slushed to lower the broken ore in the stope. Consequently, the crew on the next shift will have free access to the entire stope and can begin barring down and working immediately. Stopes at Idarado are mined until the draw becomes difficult to control. This frequently is about 76m, at which point finger raises are driven up to gob waste from the next scram and provide ventilation. A catwalk is installed to facilitate draw of the broken muck. Some stopes have been mined the full distance between levels, but this is exceptional.

When the stope is finished, draw begins immediately by slushing the first pocket empty. If the ore rills well, a large percentage can be drawn from the first few pockets. Washing and bombing from the catwalk above generally is necessary to free hang-ups and remove the muck from pillars and flat dips in the footwall. Large muck and slabs are broken in the scram as encountered without hindering other operations in the mine.

When the draw is complete, scram pillars are removed and the stope frequently is used as a waste gob hole. Many of the stopes between the 366 and 732 levels in the mine have been filled in this manner, although no attempt is made to fill particular ground. Table 6-2 shows 1969 Idarado data.

Table 6-2 1969 Idarado data—409,956t milled

Name	Operating charges per t milled, $
Labor	5.84
Supplies	3.17
Electric power	0.49
Miscellaneous	1.96
Total	11.46

Continued 6-2

Name	Development cost per meter	
	For 1, 458 meters of drift, $	For 343meters of raise(includes 335meters of doghole), $
Labor	55.51	100.85
Explosives	20.31	29.07
Timber	3.87	19.03
Indirect	127.53	228.28
Overhead	76.15	137.60
Total	283.37	514.83

Name	Typical incentive payments, $
Drifts, per m	45.93
Raises and dogholes, per m	32.81
Scram, per m	29.53
Hogging, per m^3	5.3
Stoping, per t (depending on width and tonnage)	0.68—1.57

Miscellaneous Stoping Data: 44t per man-shift in stope
22 holes per man-shift
1.3t per m of hole
0.56kg powder per t

6.2.2 Sublevel Caving

The increased application of the sublevel caving system of mining in recent years may be traced to many new technological developments, increasing labor costs, and a better understanding of the mechanics of this method. Sublevel caving has found application in both weak and strong ore bodies and may be summarized as a high-production method combining a minimum of labor and a high degree of mechanization. In its modern form it is considerably different from the methods dealt with in the literature of 30 yr ago. In some ore bodies, mining companies claim the method is comparable in cost to block caving or open sublevel stoping, with which it is most competitive.

The method is best summarized by examining Figs. 6-27 and Figs. 6-28. Basically the stoping procedure consists of driving a series of sublevels commencing at the top of the ore body. A starting vertical slot is cut and then a series of ring patterns drilled and blasted, the broken ore being drawn off after each blast. The capping must cave and follow the ore down during the drawing off procedure to localize the broken ore and to avoid serious weight problems and the possibility of dangerous airblasts. Some dilution of the ore is inevitable with the method used at present, and careful draw control must be exercised if a portion of the ore is not to be lost. Ore recovery and dilution under controlled conditions compares with that obtained from block and panel caving.

Sublevel caving is most suitable for massive and steeply inclined medium width ore bodies. Tabular ore bodies down to a minimum width of 6m have been mined by sublevel caving in Zambia

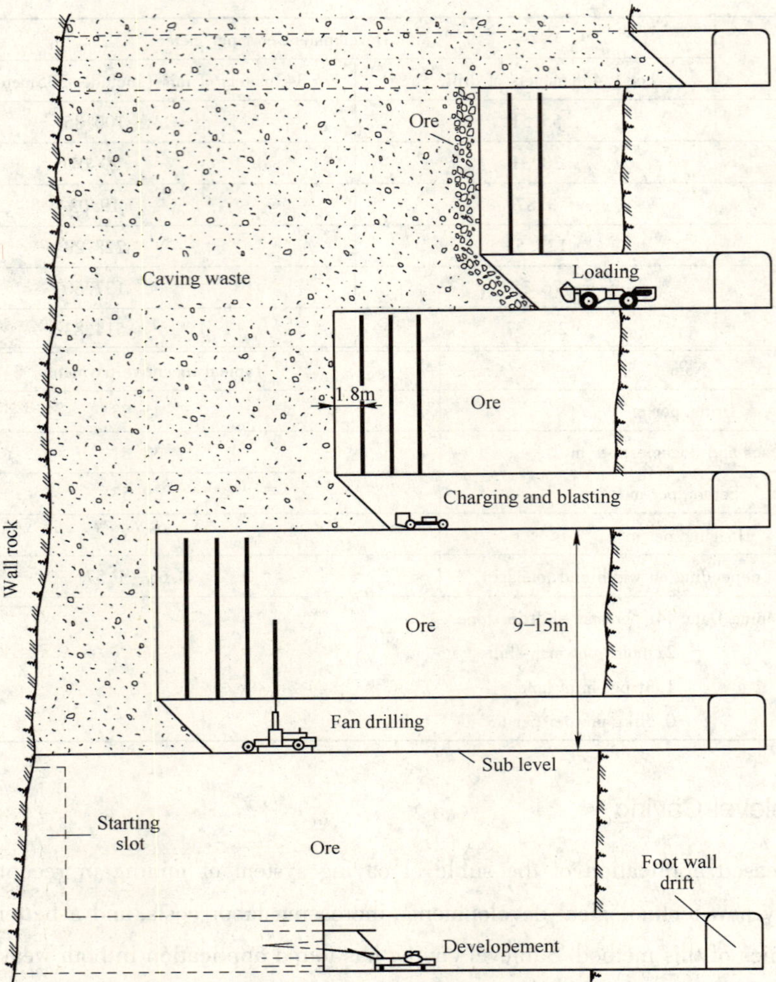

Fig. 6-27 Diagrammatic representation of a typical sublevel caving operation, showing the four principal operations

with the sublevels running parallel with the strike of the ore body.

The major requisites for this type of mining are that there should be adequate room between foot and hanging wall to drill a long-hole fan, and that the capping and hanging wall cave easily and preferably not be excessively fine relative to the size of the fractured ore. As with all caving methods, the topography must be amenable to caving, particularly with respect to surface and ground water that might flood the proposed mine workings. Sublevel caving was originally designed for weaker ore bodies. However modern developments in drilling and blasting have made the method highly desirable in many of the inherently harder and stronger ore bodies, such as the competent ore bodies of northern Sweden. Some major advantages over open stoping methods in strong ore bodies are the reduced labor requirements and no pillars to be recovered or stopes to be filled.

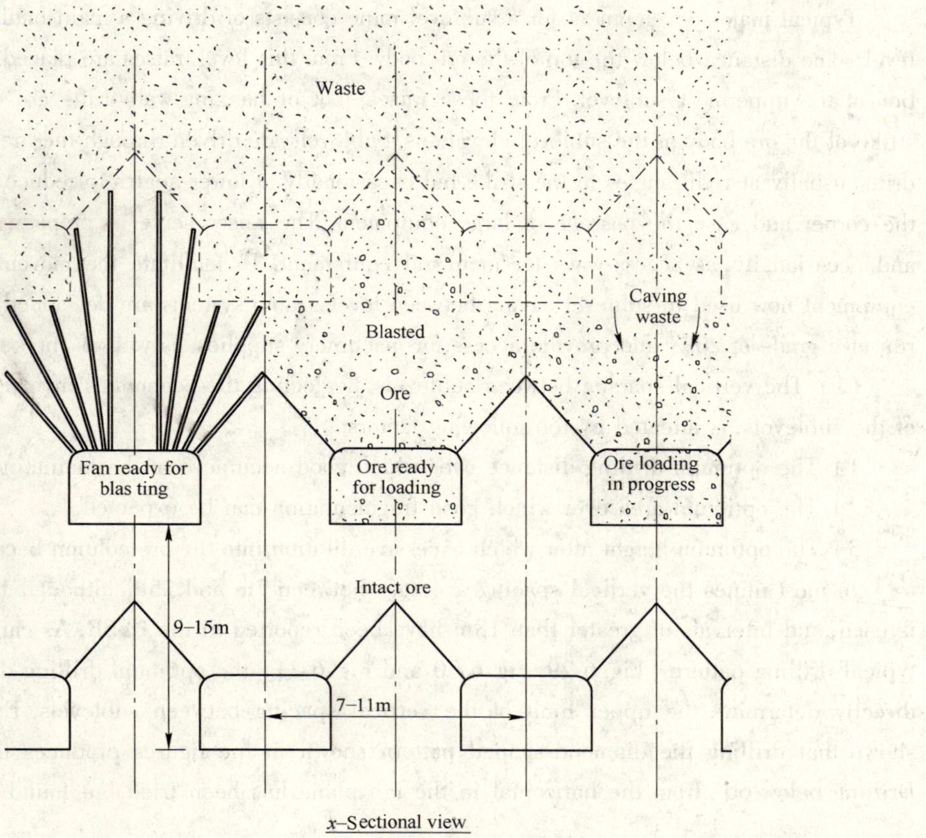

Fig. 6-28 Diagrammatic cross section of a typical sublevel caving operation showing the sublevel arrangement

Since some dilution is inevitable, a cheap method of ore beneficition must be available for any ore body in which this technique is to be used. Practice dictates that where high ore recoveries are required these will of necessity produce high dilution. Optimum operations at present show from 15% to 20% dilution with an 85% to 90% ore recovery.

6.2.2.1 Sublevel Layouts

The ultimate layouts on any sublevel caving property must be a compromise between a variety of factors. Optimum draw characteristics must be weighed against the number of sublevels and the costs of developing them, as well as the necessity for leaving adequate pillars between sublevels for ground support. Mines using the method often are continually experimenting to improve their working layouts and achieve more economical extraction. In ore bodies of variable strength and content, it may be necessary to alter stoping geometry and dimensions from one area of the mine to another, and careful examination should be made of the geology of the proposed mining areas to facilitate anticipation of changes in mining plan.

1. Major Development

Typical major development for a sublevel mine consists of driving a main haulage or collecting level some distance below the top of the ore body. From this level, raises are placed up to the elevation of the uppermost sublevel. From these raises, foot or hanging-wall drifts are driven along the strike of the ore body at the sublevel elevations. Sublevels are driven through the ore body from these drifts, usually at right angles to the strike but occasionally at lower angles to reduce the sharpness of the corner and ease the passage of large equipment. The raises serve as orepasses, return airways and, occasionally, as access-ways for men and equipment. To facilitate the movement of the large equipment now used in sublevel caving between levels, ramp systems are developed. These generally run at a grade of 20% and provide access for machinery supplies, as well as intake air.

(1) The vertical spacing between sublevels, neglecting the economics involved in the drifting of the sublevels, is affected by the following factors:

1) The optimum drilling distance over which good accuracy can be maintained.

2) The optimum area over which good fragmentation can be expected.

3) The optimum height after which excessive dilution into the ore column becomes inevitable.

In most mines the vertical spacing averages between 7m and 15m, although 18m occasionally is used, and intervals of greater than 18m have been reported in the USSR. As can be seen from a typical drilling pattern (Fig. 6-29, Fig. 6-30 and Fig. 6-31) the optimum drilling distance does not directly determine the upper limit of the vertical spacing between sublevels. Experiments have shown that drilling the diamond-shaped pattern shown in the figures produces the best results. Drilling below 60° from the horizontal in the fan plane has been tried but found ineffective, as a

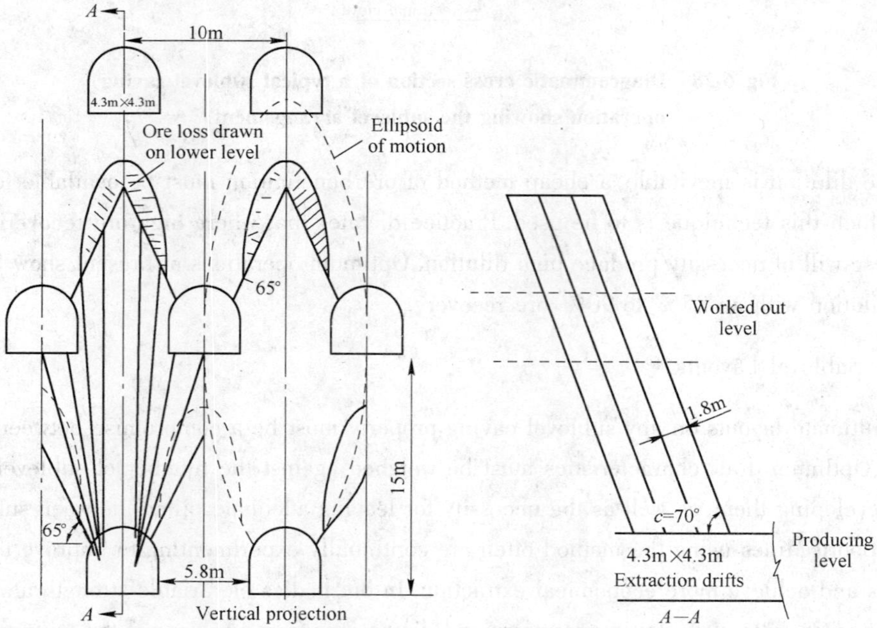

Fig. 6-29 Sublevel caving layout at Mufulira, showing expected ore loss on sides of fans. Estimated ellipse of motion shown in broken lines

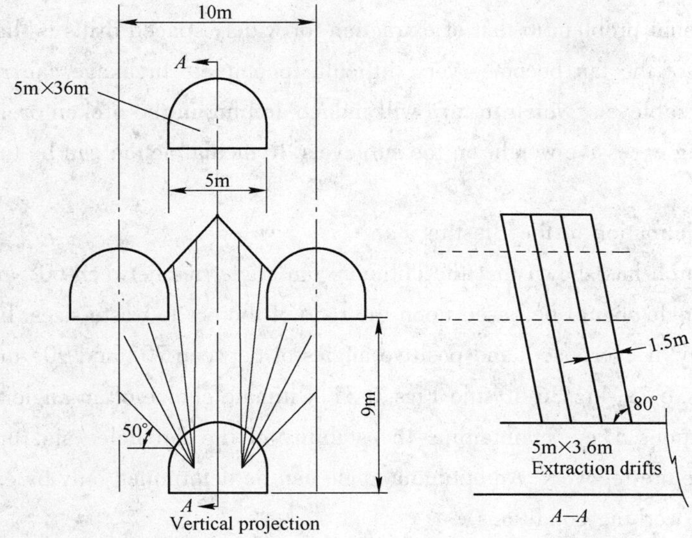

Fig. 6-30 Sublevel caving at Kiruna, Sweden. Short vertical interval between levels is necessary to provide selectivity in mining variable-grade ore

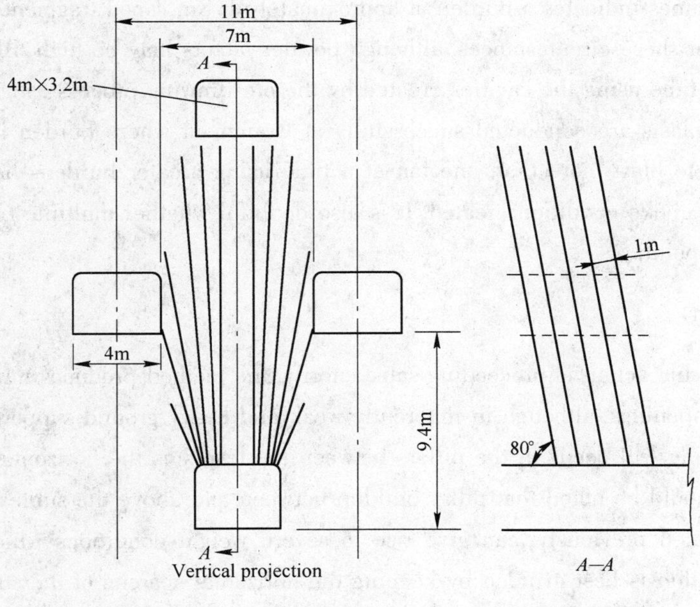

Fig. 6-31 Sublevel caving at Craigmont, Canada

choke blasting condition can occur. Typical angles run from 70° at Koskullskulle to 74° at Craigmont.

(2) Horizontal distance between sublevels. In terms of draw theory, optimum extraction could be achieved if the pillar width between sublevels equaled the width of the sublevel. This generally is not possible, for both economic and structural reasons. Actual distances between sublevels range

between 9m and 15m (centers), the minimum distance being the more preferable in terms of extraction. An additional problem to that of extraction for widely spaced drifts is that breaking ground in the outer limits of the fan becomes very difficult. Incomplete breakage can result in forming a pillar between the sublevels, which in turn will induce arching in the broken ore, thus retarding extraction and causing excessive weight on the sublevels. Remedial action can be time-consuming and expensive.

2. Angle of Inclination of the Blasting Fan

Swedish research has shown that the optimum fan angle lies between 60° and 120° and that, theoretically, the angle should be based upon the ratio of the ore to waste sizes. In practice, a backward angle is rarely, if ever, used, and positive angles of between 70° and 90° to the horizontal are most common, Figs. 6-29, Figs. 6-30 and Figs. 6-31. Ultimate choice of an angle must depend upon local mining conditions, i. e., maintaining the stability of the fan holes and the brow, combined with the percentage ore recovery. An optimum angle can be determined only by experiment with the slope under actual working conditions.

3. Fan Burden

Swedish work shows the optimum burden to be 2.1 m. Under Swedish conditions, this burden produces optimum recovery, and dilution increases if the burden is greater or less than this value. Practice in most mines indicates a burden of approximately 1.8m. Good fragmentation and recovery are achieved under these circumstances, although powder factors may be high. Blasting normally is done one fan at a time using the cavities created by the ore-drawing process as a "free face", Fig. 6-27. Double fan blasts are conducted successfully at Craigmont where burden is reduced to 1m. However attempts to blast more than one fan at a time using greater burdens have not been successful due to the choke condition created. It is also doubtful whether multiple blasts are desirable from a draw standpoint.

6.2.2.2 Support

When the caving action is proceeding satisfactorily, the method produces minimum weight upon the extraction openings, although in inherently weak and heavy ground support may be necessary. Maximum loading will occur in the pillars between the levels as the horizontal sublevel spacing is minimized. It should be noted that pillar buildup between and above the sublevels into the caved capping as mentioned previously, can give rise to severe weight conditions which are difficult to remedy. This situation is best avoided by keeping the horizontal spacing of the sublevels to a minimum, thus facilitating complete blasting between the pillars. Depending upon the vertical spacing, the uppermost sublevels are some 9 to 15 m below the caved capping and thus are rarely liable to point loading as is often experienced underneath a block cave. In addition, the mobility of the working end usually will not allow time for a significant stress buildup to develop as a result of the drawing action. Support, therefore, need not be excessive, can be temporary, and often is recoverable. In practice, support may vary from nothing, or a light bolting pattern, to heavy regular timber sets in naturally heavy ground.

Brow wear over the loading places due to the attrition of the drawing ore and the weakness of the ground has proven a serious problem in some mines and is best exemplified by conditions at Craigmont where, after much experimentation, 2.4m rebars set in concrete achieved generally good brow control. In extreme situations involving brow failure, posts and caps may have to be used, with a corresponding decrease in the width of the loading place.

6.2.2.3 Ventilation

Ventilation in the sublevel represents a problem as the drifts are single ended and no circulation of air is possible. Ventilation therefore must be introduced via ducting, the quantities required depending not only upon the drift size and number of men working in it, but also upon the machine requirements. In the case of diesel equipment, ventilation availability can be a major factor in determining the power of equipment that is practical.

6.2.2.4 Equipment

Since the sublevel caving method can be very highly mechanized, the correct selection of equipment is essential to achieve optimum mining conditions. The size and construction of intended equipment must depend upon ground conditions, as well as the geometry of the proposed system—i. e. sharpness of corners—and must represent in some cases a compromise between the increased production afforded by the larger equipment and the increased support and drift maintenance costs produced by maintaining excessively large openings. Overall, the tendency is distinctly toward trackless vehicles with heavy emphasis being placed on the self-contained load-haul-dump (LHD) units for short to medium haul distances.

1. Loading and Ore Handling

Two basic arrangements are used for ore handling in the sublevel. The first encompasses a large capacity front-end loader, such as a Scoopmobile or Atlas Copco T4G Autoloader, which combines the loading and hauling phase of the operation. These are most common in the smaller-sized sublevels (Table 6-3). The loader will tram along the sublevel and dump directly into ore passes. The alternate arrangement used in the larger sublevels requires a front-end loader in conjunction with a shuttle car, typically a 32-ton-capacity Kiruna truck. The haul distances are extremely variable and depend ultimately on local conditions. Briefly, it can be seen that shorter haul distances can be achieved by increasing the frequency of orepasses and therefore the increased production resulting from short haul distances must be measured against the cost of raising orepasses and maintaining them. The cost of constructing orepasses can be excessive in weak ground where a heavy lining is required and the effective life is short. The tendency is heavily toward trackless mining using rubber-tired vehicles for increased speed and mobility and decreased maintenance costs.

2. Drilling Requirements

Two types of drilling machines normally are required for the subleveling method.

Table 6-3 Typical stope equipment and performance

Mine	Equipment	One-way tram distance	Capacity
Kiruna, Sweden	Joy 18R loader in conjunction with Kiruna 480 truck, capacity 38t in $18m^3$ body, powered by 179kW diesel	>200m	Output, 367t per man-shift
	LHD Scoopmobile KL 560 2LD. 450; 10.7t capacity in $3.8m^3$ bucket	<200m	
Koskullskulle, Sweden	Atlas Copco T4G rubber-tired Autoloader, $0.3m^3$ bucket capacity	Peak output, up to 508t per man-shift Average Haul Figures: m 20 40 60 80 101	 t per man-shift 290 255 229 207 189
Craigmont, Canada	ST4 Scooptram used for development and stoping. Diesel engines de-rated to 78kW to minimize ventilation requirements	Up to 366m; Average, 229m	Average, 305t per man-shift

(1) A drilling jumbo to facilitate the drifting of the sublevels (Table 6-4). These are commonly two- or three-boom trackless rigs such as the three-boom Atlas Copco Boomer 131 with a single operator.

(2) A long-hole machine for drilling the fan holes. This is usually of the percussive wagon-drill type, such as the Atlas Copco Simba 22 rig or Gardner-Denver fan drill, which are capable of drilling a hole of the required length and diameter with a single operator. Drilling up to 183m of 51mm hole per shift is not uncommon using such a setup.

3. Shot Loading

Shooting is done most commonly with an ANFO-type explosive, although slurries and dynamite sometimes are sued. Machines are used for loading both types of explosives at rates in excess of 1,000kg per hr for an explosive, and 400 to 800kg per hr for the dynamite cartridges. Good fragmentation is essential in most operations, particularly where the waste is fine, as this would tend to be drawn through prematurely if the ore were coarse. At Craigmont, a powder factor of 0.4kg per t is used with the blasting agent composed of 75% Forcite in 38mm × 406mm cartridges.

6.2.2.5 Merits and Demerits of Sublevel Caving

(1) The sublevel caving method finds application in a wide variety of ore bodies. The method is especially amenable to weaker ground which might otherwise have to be mined by square-setting or top-slicing techniques. The method is, however, not limited to the weaker ores and can be used in otherwise competent ground, as at Kiruna, in Sweden. Provided the capping will cave and there is sufficient room to drill the fans, the method can be sued if desired.

Table 6-4 Typical equipment used in stope and sublevel drilling

Mine	Stope drilling	Sublevel development	Blasting details
Kiruna, Sweden	Atlas Copco Simba 22 rig	Denver Universal 3-boom drill carrier	Stope. Single operator, 225m per shift; corresponds to 2, 438t ore per man-shift
Malmberget, Sweden	Atlas Copco Simba 22 rig	Joy Serpent 3-boom hydraulic jumbo	
Craigmont, Canada	Gardner-Denver fan drills with 2 DH 123 drills. Atlas Copco Simba 26 (modified). Fan drills with 2BBC 120 drills	Gardner-Denver universal jumbos with DH 123 drills	One man can drill 2 dev. rounds per shift, occasionally 3 Stope hole dia, 51mm
Mt. Isa, Australia	Atlas Buk Ⅱ drill rig	Gardner-Denver 3-doom drill rigs fitted with DH 114mm drills and 3m feed slides	Stope drilling averages 61m per shift

(2) The method is very amenable to a high degree of mechanization combined with minimal manpower requirements. These factors contribute to making it a low-cost and therefore very desirable method in an era of rising wages and an increasing shortage of experienced miners.

(3) The method is versatile in that it permits a wide selection of mining areas, thus facilitating the blending of different-grade ores. This factor has led to a high degree of computerized control in the Swedish operations.

(4) The method may be considered to produce high tonnages, particularly using modern LHD and drilling equipment. The number of loading places can be increased to yield the required tonnages, provided the geometry of the ore body is amenable.

In considering the method in comparison with others that might be considered for a particular ore body, the following points are worth careful examination:

(1) Ore extraction may be as low as 60% but generally will not be in excess of 80% to 85%. Although these figures are low, they are comparable to other methods where it is necessary to leave pillars and where subsequent pillar extraction is difficult.

(2) Dilution is inevitable with the method, particularly in a case where the ore is coarse and the capping extremely fine. Draw-control problems may also occur where there is little apparent physical difference between the ore and the capping. Strict draw control procedures must be employed under these circumstances if dilution is to be kept to a minimum.

(3) Ventilation in the individual sublevels must be conducted using tubing.

6.2.3 Filled Stopes

6.2.3.1 General Plan

A filled stope is one in which ground support is provided by some form of waste material (Fig. 6-32). Filling, therefore, is an integral part of the stoping process. The ore body is extracted

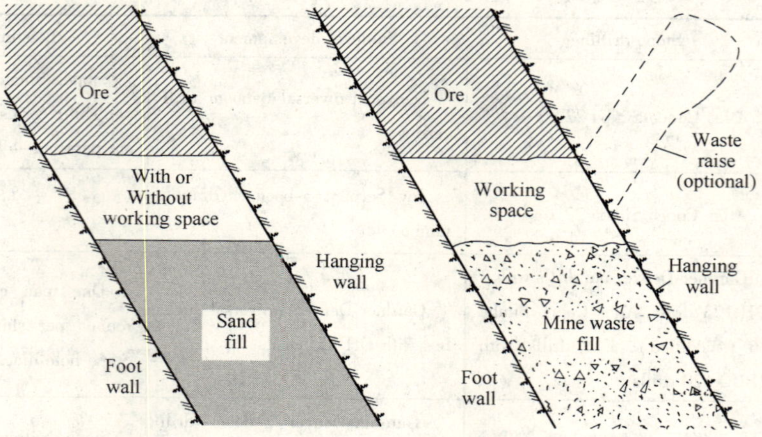

Fig. 6-32 Backfill used for ground support

in small sections which are partially or completely filled before adjacent ore is extracted. The waste material is called fill, backfill or gob. The most common forms of fill are waste rock, sand or gravel, and tailings. Timber, when used, is for temporary support of the walls or back and is not necessarily systematic.

In most filled stopes, progress is upward. Supported openings are extended through the fill for access, ventilation and ore removal. Stope characteristics, however, are variable. Development to and within the stoping area varies with the strength of the ore and wall rock and with the configuration of the ore body. The size of the section mined before filling depends on the strength of the ore and wall rock. The ore-removal system varies with the properties of the fill, equipment available and overall economics.

1. Applications and Variations

The most common applications for filled stopes are in steeply dipping ore bodies with restrictive dimensions and weak walls where complete extraction or selective mining is desirable. Filling also is applicable where ground subsidence must be prevented or controlled. Cut-and-fill stopes can be converted to methods giving various degrees of support as ground conditions change (Fig. 6-33).

Filled stope variations commonly are classified as:

(1) Resuing.

(2) Horizontal cut and fill, or flat-backs.

(3) Inclined cut-and-fill, or filled-rill.

(4) Timbered and filled.

(5) Undercut and fill.

(6) Combination methods.

2. Trends

Trends in cut-and-fill stoping include:

(1) Mechanization involving trackless equipment, raise boring for rapid stope development, and ramp access.

(2) Pneumatic backfill using stowing equipment.

(3) The use of additives in fill to increase its strength.

(4) More rapid mining of small blocks to control pressure and minimize ground movement, and eliminate combination methods in pillar recovery.

(5) The use of rockbolts, mats, mesh and shotcrete for temporary support.

(6) Continuous mining where longitudinal stope sections or a group of stopes in one block can be drilled, blasted, supported and filled concurrently.

(7) Decreasing use of soil cementing, timbered flooring over fill, hand filling and other time-consuming practices.

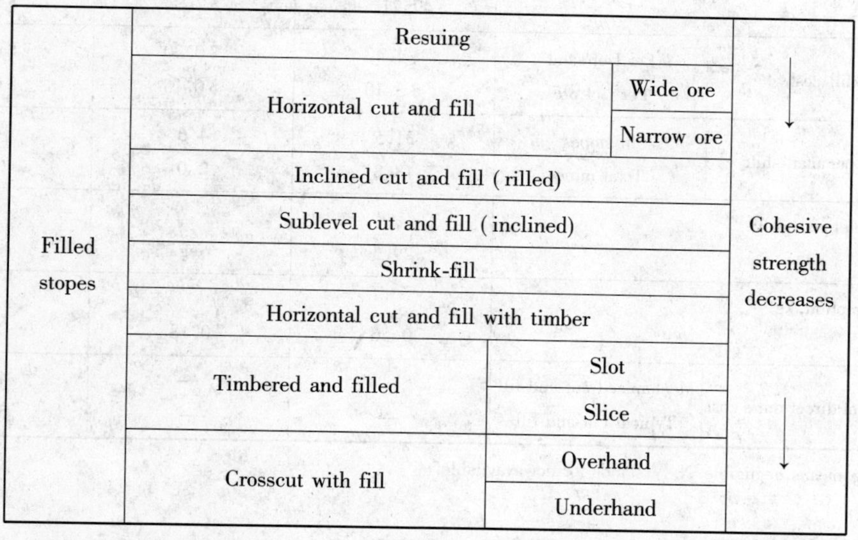

Fig. 6-33 Filled-stope variations relative to ground strength

3. Statistics

Traditional filled-stope methods have been considered relatively expensive methods with low output. However, recent improvements have made filled stopes economically more attractive. The cost of mining normally ranges between shrinkage and square-set methods. Productivity in mechanized operations approaches that of room-and-pillar mining.

Some average statistics for cut-and-fill operations are summarized in Table 6-5. These statistics include exploration, development, stoping, filling, hoisting, hauling, pumping and mine general costs.

4. Advantages and Disadvantages

Advantages of filled stoping methods include:

(1) Ground stability is maintained.

(2) Low dilution with good recoveries of ore is possible, and waste that must be mined fre-

quently can be left in a stope.

(3) Ore can be removed as it is broken, which minimizes adverse chemical reactions and provides early return on the expenditure required in mining the stope.

Inherent disadvantages to the method include:

(1) Stope production is not easily varied and stope output is irregular because of the intermittent breaking of ore as a result of the fill cycle.

(2) Backfill must be provided as required.

(3) Mining costs are increased by filling.

Table 6-5 Statistics for cut-and-fill stopes

Item		High	Low	Approximate average
Direct mine operating costs: Per t of ore		\$49.21(+)	\$2.07	\$9.65
Hydraulic backfill cost	Per t of sand	\$4.42	\$0.30	\$0.80
	Per t of ore	\$3.10	\$0.15	\$0.64
Productivity, t per man-shift	In stopes	31.5	4.6	14.3
	Total mine	18.8	2.0	6.5
Timber consumption, bd ft: Per t mined		29.5	Neg.	N. A.
Powder consumption, kg: Per t broken		0.56	0.18	0.33
Percent labor of direct mine cost	Mechanized cut-and-fill	—	—	40%
	Typical cut-and-fill	—	—	57%

Note: Neg. means negligible. N. A. stands for not avavilable.

6.2.3.2 Ground Control

Pressures from fill and wall rock become progressively greater as mining advances. This increase in pressure frequently causes serious maintenance problems in the openings serving the stopes. Pressure from fill can be partly overcome by leaving pillars of ore around mine openings. This leaves ore in hard-to-mine pillars. Pressure from the wall rock can be partly overcome by mining stopes of small dimensions. Mining and filling small stopes rapidly with dense material often delays ground movement and the accumulation of pressures from the rock masses. Pressures from weak walls in thick ore bodies or in stopes at considerable depth can be overcome by alternating transverse stopes and pillars. The pillars support the hanging wall and footwall until consolidating backfill is placed against them. The basic pattern is dependent upon the width of the ore and the strength of the rock. Transverse stopes create the problem of pillar recovery and require a second stage of mining.

Stope walls are free to converge a considerable distance before the fill is compacted sufficiently to offer any appreciable resistance. In Bureau of Mines tests in relatively narrow stopes, wall pressures approached 4134kPa and vein wall closure in excess of 508mm was recorded. Increasing the density of the fill would appear to be highly defective against such closure. A significant increase in density can be achieved through compaction or cementation or both.

Hydraulically and pneumatically placed backfills are the least compressible, provided they are properly sized and deposited. Hydraulic fill rarely compresses more than 20% and generally from 5% to 10%. This fill offers more resistance to wall closure when its moisture content is 10% to 20%. Recent operating results indicate that properly placed pneumatic fill will provide even better strength than hydraulic fill. Loose dry fill provides the least compressive strength and may shrink up to 25% in volume.

Reduction in vein wall closure facilitates mining on all sides of a filled stope. It reduces stress concentrations in unmined pillars and lessens the extent of overall ground subsidence. The use of fill controls slabbing, which prevents accelerated growth in the size of the weakened area.

6.3 Surface Mining

Fig. 6-34 shows a bird' s-eye view of an open-pit mine.

6.3.1 Drilling Machines

Fig. 6-34 A bird' s-eye view of an open-pit mine

The following discussion is limited to primary blasthole drilling as practiced in open-pit quarry and stripping operations. Although primary blastholes usually are drilled vertically, angle holes also are common, and most blasthole drill rigs, therefore, have angle drilling capability. A few rigs are designed to drill horizontal holes for overburden removal, but these are limited in number. Primary blastholes range from 51 to 432mm in dia and commonly are drilled to depths of 9 to 18m or more. Most blasthole drill rigs use either rotary, percussive or rotary-percussive drilling methods, with jet piercers used to a limited extent. Since all blasthole drills are designed for high production, they therefore feature fast drilling, rapid tramming, quick setup, automatic drill steel handling, and rapid hoisting and lowering speeds.

6.3.1.1 Rotary Blasthole Drilling

Rotary drilling is the most versatile of all drilling methods. It can be employed in very soft material when drag bits are used, and in medium to very hard rock when rolling cutters are sued. These bit types are described and illustrated later. Most rotary blastholes are shallow, from 9 to 18m, and are in the 152 to 229mm range, although the rotary method is capable of drilling 76 to 432mm holes to depths well in excess of those required for blastholes. Blasthole rigs are designed for high production and in soft material can drill 244 to 366m of hole per shift. A rolling cutter with air circulation is the most common. Drag cutters, with air or water circulation or augers for cuttings removal, can be sued to good advantage in soft material, such as overburden, soft sediments, coal, etc.

The trend in rotary blasthole drilling is to larger-diameter blastholes to take advantage of the lower costs of drilling on a volume basis and the increased efficiency of the blasting agent in larger-diameter holes. This trend has necessitated the development of heavier drill rigs, some capable of applying a 54,432kg thrust to the bit. Rotary drilling with rolling cutters now is used to drill even the hardest rocks, such as Minnesota taconite.

Rotary drills used in open-pit blasthole drilling consist of a power source, a drill string composed of a single or a connected series of hollow drill pipe, and a dill bit. The power source simultaneously applies a downward thrust and a rotary motion to the drill string. The drill pipe transmits this thrust and rotary motion to the bit and also serves as a conduit for the flushing medium. In normal drilling, the flushing medium, either air or liquid, is forced down through the drill pipe and across the face of the bit where it picks up the drill cuttings and carries them back to the surface through the space between the outside of the drill pipe and the hole wall. The drilling torque supplied by the surface power source is the energy required to fragment the rock and is consumed as the cutting elements of the bit chip and crush the rock surface. In soft material, rock disintegration is effected predominantly by a plowing action; in hard material, by indention; and in medium strength rocks, by a combination of these processes.

1. Rotary Drill Rigs

The main components of rotary blasthole rigs are the mounting, the prime mover, the rotational system, the pulldown and hoisting system, the circulation system, and the mast and rod handling system. The diesel-powered crawler-mounted drill shown in Fig. 6-35 illustrates these features.

2. Mounting

Rotary blasthole rigs are either truck or crawler mounted. Large heavy drills are always crawler mounted, as are many of the medium-sized rigs. A crawler mount can support a heavier rig than a truck mount and can move more easily on rough or inclined surfaces. Lighter rotary rigs and augers generally are truck mounted. Trucks are more mobile and have advantages if long distances and frequent moves from one site to another are required. Regardless of mounting, all drill rigs are leveled and supported by hydraulic jacks that are individually adjustable for setup on uneven ground.

3. Power Source

The prime mover for the rotary rig may be a gasoline or diesel engine or an electric motor. Although electric power is the most economical, diesel power usually is preferred when mobility and self-containment are important. Power may be transmitted mechanically or, more commonly, through hydraulic devices, and is required for the following: (1) tramming the rig from site to site; (2) operating the jacks and erecting the mast; (3) generating the pulldown force on the bit; (4) rotating the bit; (5) hoisting; and (6) operating the air compressor. The power split on an electric rig capable of drilling 311m holes is; rotation, 52 to 104kW; tramming and hoisting, 26 to 52kW; pulldown

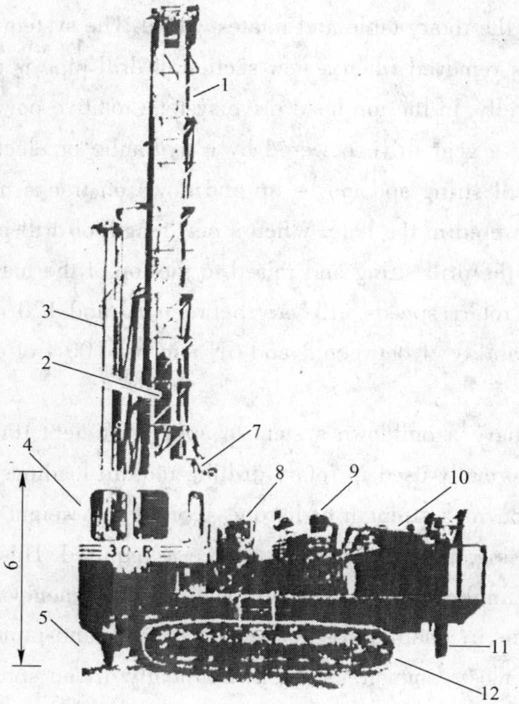

Fig. 6-35 Medium-size blasthole drill rig for holes up to 200mm in dia. Rig is adaptable for rotary and downhole percussion drilling (courtesy Bucyrus-Erie Co.)

1—Structural derrick provides height to handle rotary equipment and heavy down-to-hole tools; 2—Hydraulically powered top rotary drive with variable speed control; 3—Carousel pipe rack stores several drill pipes, for drilling deep blastholes; 4—Control panel is in enclosed cab adjacent to drill hole; 5—Dust control is by dry-type dust collector and/or water injection; 6—Low maximum height permits low-boy portability between jobs; 7—Hydraulic cylinders are provided for raising and lowering of the derrick; 8—Hydraulic oil cooler with fan; 9—Diesel engine provides power for all drilling requirements; 10—Dual staged air compressor, V-belt driven from the main engine, for rotary and down-to-hole drilling; 11—Hydraulic leveling jacks; 12—Crawler mounting

system, 15kW; air compressors, 298kW; all other systems, 30kW. Since these operations are not performed simultaneously, the rating of the power source is less than the total of the individual requirements. During drilling, the major power consumers are the air compressors and the rotational system. On small truck-mounted drills, the truck engine with power takeoff can be used as the primary power source; otherwise a separate prime mover is provided.

4. Rotational System

Systems for providing rotative power are the kelly bar system and the hydraulic or electric top head drive. The kelly bar system uses a rotary table mechanically or hydraulically driven by the prime mover. The square, hexagon, or splined kelly bar, connected to the drill pipe, slides through the matching opening on the rotary table and rotates with it. The system is rugged and dependable, but the kelly bar must be removed when a new section of drill pipe is to be added unless the drill pipe is shaped like the kelly. In the top head drive system, rotative power is applied directly to the end of the drill string by a gear drive powered by a hydraulic or electric motor. The entire drive unit is attached to the drill string and moves up and down on tracks mounted on the mast as the drill string is raised or lowered in the hole. When a new length of drill pipe is needed, the drive assembly is detached from the drill string and raised to the top of the mast to permit inserting a new length of pipe. Normally, rotary speeds will vary between 30 and 120rpm, and most drills are designed with a torque capability of between 3 and 6N · m per 100N of down force.

5. Pulldown System

All blasthole drills have a pulldown system to apply sufficient thrust to the bit to effectively fragment the rock. Bits normally used in rotary drilling require loadings of up to 1.4kN per mm of bit diameter, depending on rock strength or hardness. Since the weight of the drill string is only a small fraction of this loading, an additional down force is required. This pulldown force is obtained almost exclusively by hydraulic power applied through the movement of pistons attached to cables or chains, by pistons alone, by chain and sprockets, or by rack-and-pinion gear systems. The chain-and-sprocket system, the most common, uses a hydraulically driven sprocket gear that engages and moves two continuous roller chains attached to both sides of the top-drive motor mounting. Other systems use pistons exclusively or pistons with end sprockets that provide motion to roller chains or cables as the pistons are extended. Still another uses a driven pinion gear attached to the rotary drive unit that engages a rack mounted along each side of the mast. Small auger rigs generally use hydraulic pistons that apply thrust directly to a thrust yoke.

All pulldown systems use the weight of the drill rig to react against the down force applied to the bit. Thus, heavy drill rigs are a requirement in large-diameter blasthole drilling, and the largest weigh over 73t. In general, the drill rig will be capable of applying a down force of between 55% and 70% of its total weight. The total down force that can be applied to the bit depends not only on the total rig weight but also on how the weight is distributed on the rig. Some rigs carry ballast boxes that can be filled with sand or water to increase weight.

6. Circulation System

Because of greater penetration rates and longer cutter life, nearly all rotary blasthole drills use direct air flushing with three-cone rolling cutter bits. Water circulation is used only if water inflow into the borehole is too great to handle with air drilling methods. The normal path of circulation is from the compressor down the drill pipe to the bit, where cuttings are picked up, and back out the hole through the space between the outside of the drill pipe and the wall of the hole. The air required for bsasthole drilling is that necessary to clear the cuttings from the hole as soon as they are formed. A deficiency of air will permit the cuttings to be reground, decreasing the penetration rate. The minimum return air velocity required to efficiently clean the hole is 15m/s for such low-density materials as limestone, and 30m/s or more for such high-density materials as iron ore or where cuttings are large.

Because blastholes are relatively shallow (30m or less), air pressures of 207 to 414kPa usually are sufficient to control water inflow into the borehole. Air is supplied by one or more onboard compressors, usually powered by a separate 75 to 149kW electric or diesel prime mover. Single-stage compressors can supply air at 207 to 482kPa; a two-stage unit can deliver both low-pressure air at 207 or 414kPa and high-pressure air at 689 or 1, 723kPa. High-pressure air is required if down-the-hole hammers are to be run. During drilling, rock chips are carried to the surface, where the heavier chips normally fall out around the drill hole, and the lighter particles are either discharged into the atmosphere or collected for sampling or disposal.

If a primary dust collection system is used, a flexible shroud surrounds the drill hole and most of the smaller airborne particles are collected by a vacuum system and deposited in hoppers. If a secondary system is used, the dust-laden air is run through a cyclone separator and all but the finest particles are separated out. Another technique to reduce airborne dust is injecting a small amount of water into the air stream. This air-mist mixture agglomerates the fine dust into larger pellets, which then can be collected more easily. The collected rock chips and material around the borehole can be used later for stemming.

7. Mast and Rod-Handling System

Masts are constructed to hold one complete length of drill rod and can be built to accommodate any length up to 15m or more. Masts are hinged and connected to hydraulic cylinders. This permits the mast to be lowered, if necessary, when moving the drill rig. If the drill center of gravity is low, the move short and the terrain suitable, the mast need not be lowered when moving. If angle drilling is required, the mast can be pinned and braced in a partially upright position. Most drills are capable of angle drilling from 0 to 30° off vertical.

Most drill rigs have a rod carrier attached to the drill mast for storing and automatic changing of drill rods. Rod carriers are of either the boom type, which handles a single length of rod, or the magazine or barrel type, which stores 3 to 10 lengths of rod, or enough to drill 30 to 46m. Magazine-type rod carriers are automatically indexed to provide a loaded chamber each time the loader is swung into place.

8. Rotary Drill Bits

The two main types of bits used in rotary drilling are drag and rolling cuter. Drag bits depend on a plowing or scraping action to fragment material and are confined to drilling soft material. Although thrusts required are much lower than for rolling cutter bits, their torque requirements are much higher per unit of thrust. Drag bits typically are used in soils, soft shales, poorly cemented sandstones and other soft sediments.

The rolling cutter bit is of either two-cone or three-cone type. The latter has three bearing-mounted rolling cones equipped with steel teeth or tungsten-carbide buttons, and the two-cone, two roller cones. Bicone bits are used most often in tilted laminated formations, as they tend to keep hole deviation within acceptable limits when used with light weights. Two-and three-cone units are classed as soft, medium, hard and very hard formation bits (Fig. 6-36). Soft rock bits have large widely spaced steel teeth with the cones mounted or skewed from their true rolling attitude so as to skid when turning. The skidding promotes a gouging plowing action like that of the drag bit and aids in drilling in soft material.

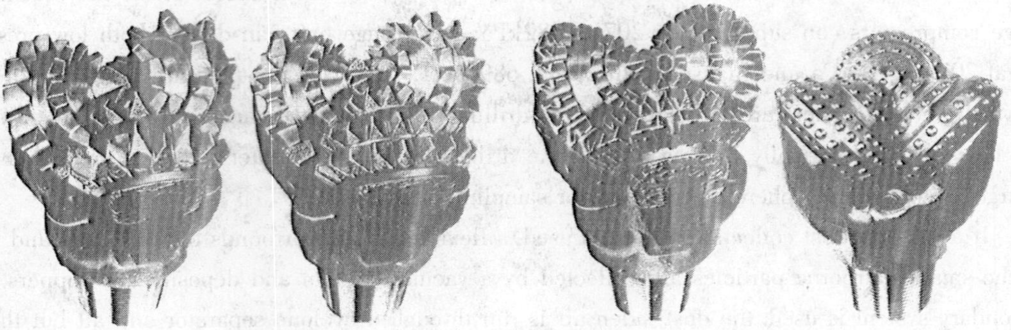

(a) (b) (c) (d)

Fig. 6-36 Tricone bits

(a) Soft-formation; (b) Medium-formation; (c) Hard-formation;
(d) Very hard-formation—tungsten-carbide-insert

Hard-formation bits generally have smaller teeth or tungsten-carbide inserts with less relief, and the gage areas are reinforced to minimize gage loss. The cones of hard-rock bits provide a true rolling action with little or no skidding. Drilling is accomplished with the rock failing under pure impaction and indentation of the cutter elements, with little or none of the scraping action of the soft-rock bits. Rolling cutter bits designed for air drilling have special built-in air passages which divert approximately 20% of the total flushing air for cooling and cleaning the cutter bearings. This air diversion is the primary difference between rolling cutter bits used with mud or water and those used with air circulation.

6.3.1.2 Percussive Blasthole Drilling

This section describes drilling blastholes for surface mining or quarrying operations by percussive methods. Percussive drills break rock by high-frequency high-energy blows applied through a chisel or button bit. The impact energy is developed by a piston that strikes the bit or drill steel and produces a high-energy stress wave that is transmitted to the bit and crushes and chips the rock when the blows are of sufficient magnitude. The bit is indexed between blows to continuously provide a fresh surface.

Percussive drills are most commonly used to drill small-diameter holes in hard rock and larger-diameter holes where heavy rotary rigs cannot be used or are not available. Percussive drilling is employed for holes from 50 to 305mm in dia to depths of 30m with surface drills, and in excess of 30m with down-the-hole units.

The primary advantages of percussive drilling are: (1) small-diameter holes can be drilled in the hardest rock; (2) percussive drill rigs are lighter, less expensive and more maneuverable because they require only a few thousand pounds (note: 1pound = 4.4453N) of down force; (3) percussive rigs can be operated by one man and can drill at any angle through a complete circle; (4) with recent developments in independent-rotation drills, percussive drilling is economical in a wide range of rock types and hole diameters.

1. Types of Percussive Drills

Percussive drills are of two major types, surface-mounted and down-the-hole. In surface-mounted drills, the percussive actuator remains outside the hole and the percussive energy is transmitted to the bit by a series of hollow drill rods. In down-the-hole drills, the percussive actuator strikes directly on the bit and follows the bit into the hole as drilling progresses.

Down-the-Hole Percussive Drills

All down-the-hole drills used for blastholes are pneumatically powered, with operating air supplied through the hollow drill steel used to support and rotate the drill. The drill provides only percussion energy to the bit and rotation is by a motor mounted on the surface drilling rig. Two typical down-the-hole drills are shown in Figs. 6-37 and Figs. 6-38.

Fig. 6-37 Down-the-hole hammer (courtesy Ingersoll-Rand, Inc.)

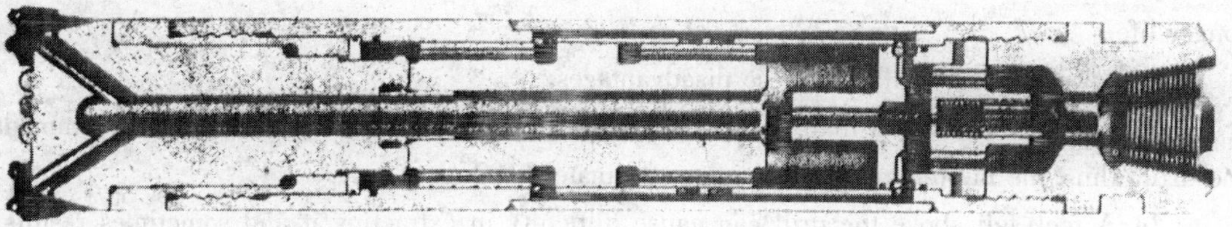

Fig. 6-38 Down-the-hole hammer (courtesy Mission Mfg. Co.)

Down-the-hole hammers are available form 83 to 254mm and commonly are used for drilling 102 to 229mm holes but with capabilities to 305mm. Down-the-hole drills commonly are used to depths of 30m or less for blastholes but can drill over 305m. The bit used must be at least 12.7mm larger than the drill diameter to provide clearance for the passage of the cuttings out of the hole. Direct air flushing is used commonly to clear the chips from the hole, although augering has been done successfully in short holes. Down-the-hole drills are operated at 689, 1,034, 1,378 or 1,723 kPa air pressure, with the higher air pressures giving the higher penetration rates. For most vertical drilling, a return air velocity of 20m/s is required to efficiently clean the hole. Down-the-hole drills generally are operated with 8.9 to 17.8kN of thrust. Additional thrust does not necessarily increase penetration rate but may increase bit wear and chance of bit failure, and eventually stall the drill. Down-the-hole drills are rotated at 10 to 100 rpm, depending on bit size, air pressure, hardness and abrasiveness of the rock. Table 6-6 lists the specifications of some typical down-the-hole hammers.

Table 6-6 Down-the-hole drill specifications

Drill specifications	Hole diameter		
	102mm	152mm	203mm
Overall diameter, mm	82.55	133.35	155.575
Tool weight, kg	41—45	82—91	136—159
Tool length, mm	1,016—1,143	1,016—1,524	1,270—1,651
Piston diameter, mm	63.5	114.3	127
Operating pressure, kPa	689—1,772.5	689—1,772.5	689—1,772.5
Air consumption, m^3/s	0.047—0.236	0.142—0.283	0.236—0.755

Down-the-hole drills generally are recognized as having the following advantages over surface-mounted percussive drills.

(1) Higher penetration rates in deep holes due to the fact that the energy loss between the drill and the bit is held to a constant minimum because the percussive energy is not transmitted through drill steel or couplings. Thus, the down-the-hole drill delivers full striking power directly to the bit regardless of hole depth. This is a substantial advantage since it is estimated that a 50% loss in drill rate results after only five or six steel and couplings additions.

(2) The exhaust air from the down-the-hole hammer provides a portion of air required for chip removal.

(3) The noise generated by operation dampened as soon as the drill enters the hole.

(4) Because no impact energy is transmitted through them, the drill rod and couplings have a longer life.

The down-the-hole drill has these disadvantages:

(1) The hole diameters that can be drilled are limited to those large enough to accommodate the drill. Thus, the minimum size hole is approximately 102mm.

(2) A rock fall above the drill can cause difficulty in extracting it and sometimes results in the loss of the drill in the hole. To minimize this risk, down-the-hole drills should be used with caution in blocky caving ground.

(3) In down-the-hole drilling, the cylinder bore necessarily must be smaller than the drill bit. This requirement is a disadvantage compared to surface-mounted drills since it limits the amount of energy that can be applied to the bit. Surface-mounted drills can, and usually do, use bits smaller than the cylinder bore for optimum drilling.

Although there are comparatively few manufacturers of down-the-hole drills, steady advances are being made in this type of drilling. The trend is to increase the operating air pressure to achieve faster cycling with about the same energy per blow. This faster cycling gives higher penetration rates without unduly disturbing the hole walls or adding extra stresses on the drill bit.

The advantages of down-the-hole drilling in deep holes and the in-the-hole noise damping will tend to increase future use of this type of drilling.

Surface-Mounted Percussive Drills

Surface-mounted percussive drills are the most widely used percussive drilling devices. The percussive actuator is mounted on the drilling rig and the impact energy is transmitted through the drill rods to the bit. They usually are pneumatically powered and generally use air for flushing the cuttings out of the hole. Surface-mounted drills are used to drill 50 to 152mm holes. They deliver 1,200 to 2,500 blows per min and rotate the bit at between 100 and 400rpm. Because of the loss of energy in the drill string in deep holes, this drill commonly is used to depths of 30m or less but has been employed to 122m in special situations. In addition to air-powered drills, a limited number of small gasoline-powered hand-held drills and larger-diameter hydraulically powered drills are in use. Hydraulic percussive drills have been recently introduced in the United States and England, but due to lack of data it is difficult to predict their effect on the state of the art of percussive drilling at this time.

2. Percussive Drilling Tools

Drill Bits

The drill bits used in percussive drilling are of two major types: chisel and button. They are further subdivided into different designs to better adapt to all kinds of percussive drilling, as listed below:

(1) Cross bits—four sintered tungsten-carbide inserts set at right angles.

(2) X-bit—four sintered tungsten-carbide inserts set at angles forming an X (Fig. 6-39).

(3) Chisel bits—one sintered tungsten-carbide insert.

(4) Cross bits—four sintered tungsten-carbide inserts with a special rake angle designed for rotary-percussive drills.

(5) Button bits—flat bottom bit with a number of cylindrical tungsten-carbide inserts, more common in larger sizes 50 to 165mm (Fig. 6-40).

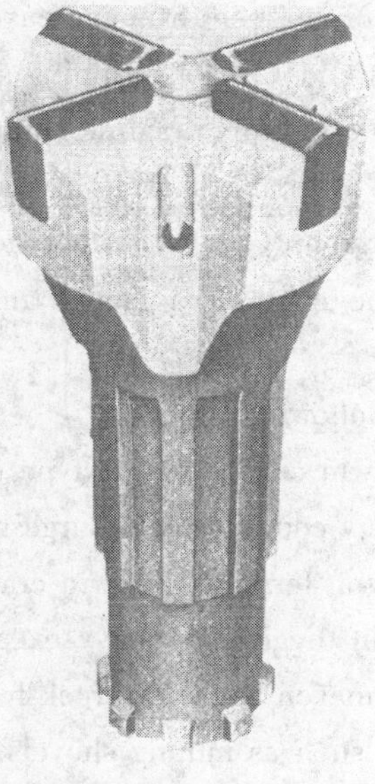

Fig. 6-39 Tungsten-carbide X-bit for down-the-hole drilling (courtesy Kenna-metal Inc.)

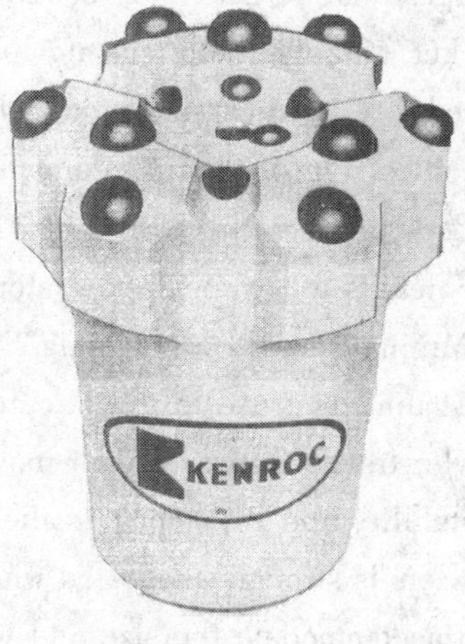

Fig. 6-40 Tungsten-carbide button bit for surface-mounted drills (courtesy Kenna-metal Inc.)

Drill Rods

The drill rods used in percussive drilling are of two major types corresponding to those used with surface-mounted and down-the-hole drills. Drill rods used with surface-mounted drills are of heavy construction to withstand the transfer of impact energy from the drill to the bit. These rods are approximately 25 to 50mm in dia with a hexagonal or round cross section. Coupled rods used with drifters are available in a range of lengths (2.4 to 3.7m), diameters and types of thread. Rods used with down-the-hole drills are lighter since they need not transmit impact energy to the bit. These rods are relatively thin walled tubing, threaded on each end, and of sufficient strength to resist the torque and thrust applied to the drill.

6.3.2 Excavation and Loading

6.3.2.1 Shovels, Backhoes, Draglines, and Clamshells

The nature and characteristics of rock and rock derivatives forming the earth's crust vary so greatly that it is impossible to use the same numerical factors on all proposed excavating projects. Similarities often occur, so historic data often is used as a background for performance of one similar job to another.

Power shovels, draglines, clamshells and hoes are applied to varying and extremely broad ranges of work. They are designed to accommodate a wide range of working conditions.

In mines, the opportunity to select the shovel or dragline most suitable for each job is present under most conditions.

The various types of traction facilitate serving as a means of moving the machines and comprise the mounting upon which the entire structures of the machines rest.

Size of excavating equipment has increased almost incredibly to meet ever increasing requirements. Dipper capacities have reached $19m^3$ for mining shovels and 138 for stripping shovels. The world's largest mobile machine, recently placed in operation, is a walking dragline equipped with a $168m^3$ bucket. Manufacturers see no limitations on the size of stripping shovels or walking draglines. Availability of truck sizes to match the mining shovels is the only major obstacle to increasing these capacities. Physical and economic considerations are the controlling factors in equipment size.

Major details for cyclical excavators are presented in the following:

(1) Mining Shovels——Mining shovels are mounted on twin crawlers. This type of mounting finds almost universal use in that it can serve under almost every condition. The large bearing area of the crawler tracks insures proper movement and stability in soft terrain. The twin crawlers afford excellent stability and resistance to the shock loads involved in digging in heavy excavation. The use of crawlers is strongly suggested when the working floor is uneven and sharp rock fragments occur, and to accommodate the size and weight involved in units such as mining shovels. Mining requirements for shovels tend to minimize frequent and high speed movement of the machines over a widely separated series of operations. Fig. 6-41 shows an electric shovel loading trucks.

Fig. 6-41 An electric shovel loading trucks

(2) Crawler-Mounted Draglines——There are many applications for this type of front-end equipment in mining and stripping operations. Fig. 6-42 shows a twin-crawler diesel-driven unit as a basic machine mounting a dragline front end. Limits as to the front-end combinations of boom length and bucket size exist for these units. Quick response by these short-coupled machines permits them to be fast-acting. The ability of these units to work effectively in tight areas and to traverse rough terrain makes them desirable excavators on many mining jobs.

Fig. 6-42 Diesel-driven twin-crawler dragline

(3) Backhoes——The backhoe customarily is a secondary tool in surface mining. It is convenient for scraping off and cleaning adhering overlying material from surfaces. It frequently is used to dig down into cracks or crevices to remove material lodged in these out-of-the-way places; Fig. 6-43 shows a hoe digging a ditch. Table 6-7 shows general specifications of hoes with their various capacity and operating ranges.

(4) Clamshells——Fig. 6-44 shows the various components comprising the base, crane and

bucket used in clamshell work. A tagline is not shown. It is used to position the bucket as it is lowered to its digging position, and to keep the bucket from spinning. Clamshells are especially suited to vertical lifting of loose materials from one elevation to another. Various job functions may call for removal of material from below the level upon which the machine rests or require carrying material above the machine. In operation, the clamshell bucket is dumped by releasing the closing line and holding it with the lowering rope.

Fig. 6-43 Hoe digging a ditch

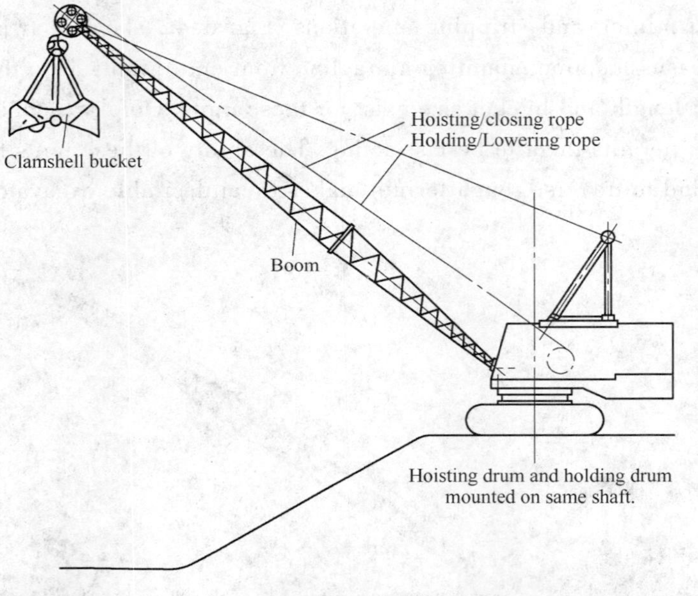

Fig. 6-44 Clamshell nomenclature

Table 6-7 Sizes and specifications—hoes with standard crawlers

Capacity range, m^3	Boom length, mm	Swing radius, mm	Max. digging depth, mm	Reach, mm	Standard crawlers		Track width, mm
					Length, mm	Width, mm	
0.19—0.57	5,334	3,048	5,563	8,992	3,334	2,629	508
0.57—0.96	5,791	3,353	6,096	9,982	3,658	2,845	660
0.67—0.96	6,706	3,683	7,163	11,430	3,689	2,946	660
0.67—1.34	7,315	3,988	7,544	12,268	3,994	3,454	762
1.34—1.72	8,534	4,724	8,915	14,249	4,140	3,505	838
1.91—2.68	9,449	5,105	9,677	15,545	4,851	3,886	838
1.91—3.25	10,363	5,791	11,048	17,069	5,118	4,191	914

6.3.2.2 Front-end Loaders and Scrappers

1. Front-End Loaders

Basically, two types of front-end loaders are used in mining. The track-type unit (Fig. 6-45) is essentially an excavating tool and is used where this alone is the aim or where stability and/or flotation is required. Use of track loaders is limited in mot areas of the world, except for European and Japanese nonmetal mines where weather and ground conditions make them desirable tools. Track-type loaders are chosen over wheel loaders for special situations, such as high digging requirements, extremely soft footing or very rough terrain requiring a high degree of stability.

The wheel loader (Fig. 6-46), on the other hand, is basically a high-speed loading tool and is used for pure loading and where a high degree of versatility and mobility is required. Wheel loaders are popular worldwide, and their use continues to increase rapidly. As wheel loaders have developed both in durability and size, their use has spread from stockpile and other loose material handling to power-shovel-type work in primary loading of blasted rock and ore, and unconsolidated materials in the bank state.

Fig. 6-45 Track-type loader

Fig. 6-46 Wheel loader

Front-End Loader Performance

Output is the product of the mass of material delivered per cycle times the cycles per hour. Cycle times vary in accordance with job conditions and loader stability. A miscalculation in either area will be multiplied 100 times or more in the course of an hour' s operation. The following tabulation makes readily apparent the effect of 0.1 min per cycle on an hourly basis.

Cycle time, min	Cycles per hr	Cycle time, min	Cycles per hr	Cycle time, min	Cycles per hr
0.3	200	0.5	120	0.7	86

Material delivered per cycle reflects bucket size, carry characteristics (carry factor) and weight of the material. A common error is assuming that a full SAE rated load (common bucket-volume description) will be delivered each cycle. In all cases, the density of the material in the loose state should be determined in t/m^3. Weighing a known sample volume by some means is recommended.

$$\text{Rated (maximum) hourly production (t)} = \text{rated bucket size } (m^3) \times \text{carry factor} \times \frac{\text{material wt}(t/m^3)}{0.9072} \times \text{cycles per hr} \quad (6\text{-}1)$$

For longer periods, an overall efficiency factor also should be applied.

Bucket Size

Buckets generally are rated on the basis of physical dimensions and an assumed 2 : 1 angle of repose of the material when the bucket is oriented so that a line drawn from the uppermost points of the cutting edge and spill plate would be horizontal (Fig. 6-47). Struck rating is arrived at in the same manner, except that the angle of repose of the material is 0 : 0. Struck capacity in relation to rated will vary slightly with bucket configuration. However, 85% of rated capacity is a rule-of-thumb approach.

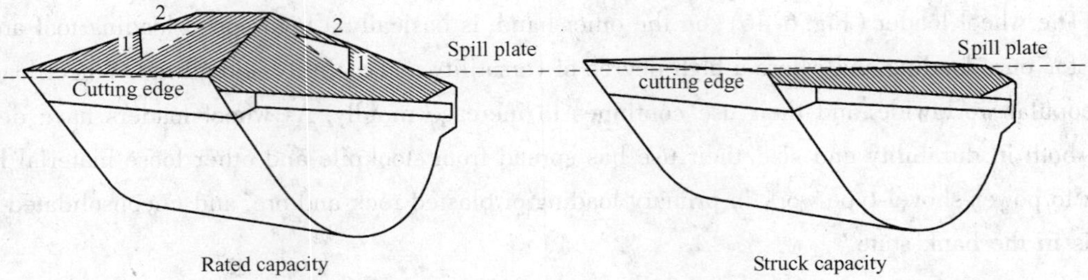

Fig. 6-47 Bucket capacities on struck and rated bases

Material Carrying Characteristics

Depending upon the characteristics of the material involved, volume actually delivered will vary between 75% and 100% (over 100% on rare occasions) of the rated capacity of any particular loader bucket. Characteristics of materials are primarily a result of the size and uniformity of the individual particles and moisture content. Clean, dry, uniform materials load the easiest but have the poorest carrying characteristics. As grain size and angularity increase, the materials become more difficult to load but carry better. Moisture content has practically no effect on materials over 3mm. Under 3mm, carrying characteristics improve rapidly with the addition of moisture up to the point the material starts to flow. The load and carrying characteristics of mixed materials are similar to those of materials under 3mm.

The following indicates the approximate volume of loose material, as a percent of rated bucket capacity, which actually will be delivered per cycle.

Loose material	Dirt or granular	Mixed	3 to 19mm	19 to 152mm	152mm and over	Slabs and blocks
Carry factor, % of rated bucket capacity	95—100 +	95—100 +	85	90—95	85—90	75—85

Loader Cycle Times

Loader cycle time is made up of load, haul, dump and return. In any stockpile or bank work, these maneuvers require four reversals of direction. In loading trucks from a stockpile, the distance traveled is held to a minimum. Depending on the operator, material, truck size and underfoot conditions, cycle time segments will vary upward from the following:

	Load	Haul	Dump	Return	Total
Cycle time, min	0.06	0.15	0.05	0.14	0.40

On other than truck-loading applications, extended hauls usually are involved. Regardless of haul length, the turns and reversals of truck loading or basic maneuvers must be completed.

Costs and Productivity

Production will vary almost directly with machine size but cost per metric ton generally declines slightly as machine size is increased.

Bucket and Loader Size Selection

Selecting the proper-sized loader involves the following steps:

(1) Determine the production rate required for a unit of time, e. g., per minute, hour, shift, etc.

(2) From cycle-time calculations, determine the adequate bucket size to meet requirements after considering carry factor and job efficiency factor.

(3) From manufacturer' s specifications and recommendations, match bucket size selected to proper loader size.

Bucket Type

There are two basic bucket types: with straight cutting edge for maximum versatility, cleanup and loading of loose graded materials, and with the pointed, or "vee", cutting edge for better penetration in blasted rock piles. Both styles can be equipped with bucket teeth for special "tight-material" situations.

Counterweights

Counterweights sometimes are required to achieve better machine stability, increase "breakout" capability, or increase the "operating-load" capability. "Safe operating load" is described by SAE as 50% or less of tipping load on a hard, smooth, level surface at speeds of 6.4km/h or less. The counterweight can be iron fastened at proper places on the machine or liquid or dry tire ballast. Both offer certain advantages and disadvantages. Personal preference and experience usually dictate.

Equipping the Wheel Loader—Tires

"Traction" and "Flotation" are two distinct and opposing factors. Traction is the friction of a body, such as a tire, on the surface on which it moves. To increase traction:

(1) For a given tire and inflation pressure, increase the weight or load on the tire.

(2) For a given load, increase the length, not width, of the tires on the ground. A larger diameter tire or decreasing inflation pressure will increase length of ground contact.

Flotation is the ability of an object, such as a wheel, to float, or "stay on top of", a surface. To increase flotation:

(1) For a given tire and inflation pressure, decrease the load, or weight on the tire.

(2) For a given load, increase the length or width of the tire on the ground. A larger diameter tire, a wider tire or decreased inflation pressure will increase flotation. Along with these measures,

increasing weight generally improves both traction and flotation in sand.

Tread designs affect traction and flotation in various materials. Aggressive tread designs, such as traction treads, produce additional pulling effort over and above traction due to friction, whenever the surface can be penetrated by the lugs and minimum slippage occurs, by taking advantage of the shear strength of the material. When slippage occurs, this tread "digs in" faster than nonaggressive treads, resulting in increased rolling resistance and less pulling ability.

Nonaggressive treads, such as sand or rock types, produce more tractive effort on loose materials with a low shear strength, as they apply less torque to the individual material particles.

For extremely soft underfoot conditions, a traction tread generally will give best results. For most mining conditions, the rock-lug tread results in greater tire life and lowest cost.

For shot-rock loading, extra-deep-tread rock tires generally provide the lowest cost per unit of production or time interval, even though higher priced. A wider tire also results in lower unit cost. As a rule-of-thumb, more rubber on the ground, within limits, provides the best results in abrasive or tire-cutting situations.

Special large balloon, or "airplane type", tires often give the best results in deep sand.

2. Scrapers

Scrapers used in surface mining are of two basic types: wheel tractor-scrapers and crawler-tractor-towed units. Most work in recent years has been done with wheel-type machines because of greater distance range, speed and productivity. Tractor-towed units generally are found at small stripping operations with short haul distances and soft underfoot conditions and where use is short-term or periodic. Wheel tractor-scrapers find their greatest use in overburden stripping. Less wide-spread is loading and hauling the mineral product to a processing plant. Other uses, secondary but vital to the operation, are dam, dike and road construction.

Wheel tractor-scrapers are designed and built in four basic configurations, each best fitted to a general set of conditions:

Three-Axle. Designed for stability and long-distance haulage at high speed on good, near-level roads (Fig. 6-48).

Fig. 6-48 Three-axle wheel tractor-scraper

Two-Axle. Designed for the bulk of conditions such as average haul distances, moderate to poor grades and underfoot conditions (Fig. 6-49).

Tandem-powered. Available in both three-and two-axle versions. The three-axle unit generally

Fig. 6-49 Two-axle wheel tractor-scraper

is high capacity for larger-volume long-distance hauling on moderate grades and moderate to good roads; two-axle, for the severest of conditions—steep grades and poor underfoot conditions, average to poor roads and short to average haul distances. They can self-load under favorable conditions (Fig. 6-50).

Fig. 6-50 Tandem-powered scraper

Elevating Scrapers. Designed for short hauls, a wide range of materials and moderate conditions; entirely self-loading (Fig. 6-51).

Fig. 6-51 Elevating scraper

Scraper Applications

Scrapers find their greatest use in unconsolidated materials that require little or no loosening before loading. However, they are finding increasing application in loading and hauling rock and ores that require ripping or blasting, or preblasting and ripping.

Users have a choice of type of equipment for a surface mining operation. Where material can be cast short distances to waste areas, as in coal stripping, large shovels and draglines generally offer the greatest productivity and most favorable costs, particularly at area-type operations. In the smaller contour mines, more mobile equipment, such as large tractor dozers, large wheel loaders and smaller crawler-mounted shovels and draglines, provide the more favorable overall performance and costs.

Where conditions, such as pit depth and irregularity, preclude casting, material most often must be loaded and hauled. Available systems include hydraulicking, excavator-belt combinations, loaders and trucks, or scrapers and pushers. The first two are limited to operations involving high volumes of loose material. The latter two provide maximum flexibility and find more extensive use. Scrapers provide the lowest loading cost and comparable hauling costs for short to medium distances. Loaders and trucks are the more favorable system for hard blasted rock and long haul distances. Detailed studies and analyses are needed to provide data for the proper choice.

Productivity and Costs

For best results, a comprehensive cost comparison should be made. Several manufacturers provide computer services for guidance in proper machine selection and requirements. For a given machine and set of conditions an optimum load time can be developed by plotting a simple load-growth curve. Optimum load time also will vary with the pusher power used.

A wide variety of types and sizes of scrapers are available. Equipment manufacturers and their representatives can provide valuable service in the forms of data and assistance in selection of size, type and number of machines and supporting equipment.

6.3.3 Open-pit Copper Mining

Besides the basic mine-planning considerations for virtually all open pit mining operations, there are three specific additional factors which require careful consideration in open-pit copper-mine planning and which are not matters of concern at most other metallic or nonmetallic open-pit mines, and therefore are not included in their normal mine planning. These are:

(1) The use of waste leaching at many copper mines, and the effect that leach copper recovery from this waste has on the ordinary milling ore reserves. This additional recovery affects not only the pit limits but also the breakeven ore grades for differentiating between mill or other process-plant ore feed, and the leachable or barren wastes which must be stripped to uncover the mill ore.

(2) The frequently significant economic influence that the minor metals (including molybdenum, gold and silver), which usually are associated with porphyry copper ores have on the ore reserves, the break-even cutoff ore grades determined for mill ore and waste, and the consequent mine planning. Usually, the simplest way of including the value of these byproducts in mine planning work is to convert their recoverable salable market values to a copper-equivalent figure, and then to base the mine planning calculations on the sum of the copper and "copper-equivalent" values. This practice requires periodic forecasting of metal prices and equally periodic revisions to the ore reserve calculations and mining plans.

(3) The normally variable contents of both nonsulfide and sulfide copper minerals in the dif-

ferent sectors or zones of the ore body. Depending on the assay grades, metallurgical responses and the physical locations of these zones, dual recovery systems may be required to achieve optimum economic metal recoveries from some zones and optimum mine life.

Further discussion of the influences of these three unique factors that can significantly affect open-pit copper-mine planning is presented in the following.

6.3.3.1 Effect of Leach Copper on Ore Reserves

The tonnage and value of precipitate, cement or electrowon copper recovered from leaching operations on copper-bearing stripped waste or below cutoff-grade mill feed has been increasing steadily and this trend is expected to continue or even increase. In 1970, production from these sources amounted to about 15% of total U. S. copper production, equivalent to about 234, 055t of copper. Recoverable values in copper-bearing overburden which is economically leachable, and must be stripped to make milling ore available, therefore should be considered when calculating ore reserves, in developing short-range mine plans and in determining ultimate open-pit limits.

When mining reaches the designed ultimate pit-limit increment, profits become marginal and the operation then would begin to close down. In the last pit-limit increment the profit obtainable from mining and milling the mill ore in that increment should equal the cost of the stripping necessary to uncover the ore in that increment. If the stripping is barren waste which must be dumped, it becomes an ordinary cost item without credits. However, if the waste contains economically recoverable leachable values, the net present value of the revenue from leaching this incremental material should be considered as a credit against the cost of this stripping. The resulting lowering of the net stripping cost enables mill ore in that pit increment to carry a proportionately higher stripping ratio, and thus increases the ore reserve and, in effect, adds one or more pit increments inside a "new" recalculated ultimate pit limit.

The determination of the net present value of the leach copper recoverable from waste must be based both on estimated leach-copper recovery rates, which ordinarily decrease as leaching continues, and on total copper recovery over the expected life of the waste-leaching operation. Even though there has been and continues to be much research into these two important factors, absolute figures on either recovery rates or total recovery are unavailable or unreliable for general use, and experienced judgment must be relied upon for establishing the basic criteria for these leach-copper recovery calculations. Leaching data from those with experience at copper mines that have been practicing large-scale leaching of waste with various mineralogical consists over extended periods provide the best source of information to project leach-copper recoveries for new copper mines being planned and for future dump-leaching operations at existing mines.

Because of the general unavailability of specific data for dump-leaching recoveries from waste or below-mill-grade copper-sulfide ores, the influence and economic effect of copper recovery from dump leaching is ignored by many engineers when estimating milling-ore reserves and determining pit limits. However, this practice can result in a significant underestimaton of a copper property' s potential if it includes major tonnages of readily recoverable values in its overburden. In those cases where copper oxide and chalcocite mineralization is present, adequately scaled test work can be done to provide the necessary and reasonably reliable data needed for leach-recovery calculations.

On a long-life copper property operating with a fixed capacity concentrator on relatively high-grade ore, and with no thought being given to expansions, it ordinarily is unnecessary to reestimate ore reserves with a high degree of precision. In these cases, the margins between copper selling prices and total operating costs are sufficient to obviate the need for frequent redeterminations of pit limits. However, if consideration is being given to increasing the property' s productive capacity, mining plans showing the grades of ore that will be milled annually before and after the proposed expansion must be developed to determine the economics of alternate expansion plans, and these mining plans should include forecasted leach-copper recovery from waste. The reasons for doing this are that in estimating ore reserves within a break-even pit-limit increment, the tonnage of material near the pit limit becomes ore only when the plant has been essentially depreciated and theoretically would net no return on investment, all when considering further leach copper recovery from waste. Such leaching may continue to be economic even after normal mining ceases.

In addition to the effect of leach-copper recovery on the pit-limit increment, it also can affect the mill ore reserves by making it more profitable to raise the mill cutoff grade and dump-leach more material below that grade. The mill cutoff grade normally is the grade at which revenue from the recovered mill copper plus other minor byproduct recovered metals will merely equal total production costs plus a minimum acceptable profit. From some ore bodies, a block of that grade can be sent to a leach dump and, although the minor metals or byproducts may not be recovered, the value of the leach copper recovered could exceed its production cost. Absolute adherence to this type of analysis ordinarily will result in a shorter mine life, and may result in a decrease in total mine copper production but at a higher net present value and/or with higher total profits. Again, in assessing the relative profitability and the risk in sending a block of ore to the mill or to waste leaching, the relative accuracy of estimating waste-leaching rates and recovery against the usually well-known concentrator recovery must be considered.

6.3.3.2 Influence of Minor Metals or Byproducts on Pit Planning and Ore Reserve Determinations

At many open-pit copper mines, the values of byproducts, such as gold, silver, molybdenum and a variety of other minor metals, is of major importance. The viability of many of the mines which have been brought into production recently is heavily dependent on the value of these minor metals and byproducts.

These minor metal or byproduct credits converted to copper equivalents are additive to the copper grade used in the calculation of break-even stripping ratios for determining pit-limit increments and in the selection of mill-grade cutoff. They also affect metallurgical process selection, operating costs and capital investment.

Whenever such minor metals or byproducts are economically recoverable, the use of copper equivalents in calculating the break-even strip ratio will increase the ore reserve and can significantly affect the initial pit plans and production rates, and probable later modifications to facilities, production rates and replacement capital.

As an example, assume an ore body with an 0.4% copper grade has a break-even stripping ratio of 3 : 1 based on copper grade only. Of the byproduct values were equal to 10% of the copper content, the incremental value of the byproducts above incremental cost could be $0.25 to $0.30 per metric ton. Thus, a stripping ratio of 4 : 1 would become economic, compared to 3 : 1 for the 0.4% copper grade.

For calculating mill-grade cutoff, including minor metal or byproduct values, the concept which should be used is: If the incremental revenue from copper and from byproduct sales exceeds the incremental cost of sending the material to the mill and the cost of further processing and marketing the products, the material becomes mill ore. The normal distribution of values in a typical copper deposit is such that large tonnages may be added to the reserves with but minor decreases in the copper cutoff grade. The presence of byproducts reduces the cutoff grade required from copper alone and thus is significant in influencing the stripping ratio.

6.3.3.3 Classification of Sulfide and Nonsulfide Material

Most copper porphyry open pits contain ore zones that have different metallurgical characteristics. The close proximity of copper ore bodies to the surface also makes it likely that the upper portion of the ore body will be partially oxidized and probably enriched.

The copper-sulfide minerals and byproducts are recovered by flotation, whereas the copper-oxide minerals normally are best recovered by dump leaching. In some cases, mined sulfide and nonsulfide ores are treated in a dual process which includes both flotation and leaching. For the less frequent situations when the copper is predominantly in silicate minerals, even more complex leaching plants are required, usually followed by electrowinning to produce “cathode” copper.

The development of both short-and long-range mining plans and pit designs should provide for the amounts and types of ore to be mined in each given time period so that the tonnages, grades and blends of mine ores will result in maximum economic return. For this purpose, it is essential to complete detailed studies of the geology and mineralogy of each part of the ore body and its metallurgical response. The economic strip ratio and grade cutoff can then be calculated for each portion of the mine, considering the recoveries and mining and treatment costs for each ore type.

6.3.3.4 Production, Operating and Cost Data

Production, operating and cost data were provided by 36 open-pit copper mines in the United States, Canada, South America, Europe, the Australian area and Africa. These detailed data and costs have been tabulated in Tables 6-8 through 6-17 of this section.

In most cases, the data represent 1971 results, with a few properties averaging their results of the last several years, as noted in the tables. With three states, four Canadian provinces and ten additional foreign countries represented in the tables, and each with its own particular operating and accounting methods, the costs and production data should not be assumed as absolutely comparative but can be considered for reasonably close comparison and estimating purposes.

Table 6-8 Production and actual costs for open-pit copper mines

Mine	Productivity	Actual cost per t of material handled, $									Explanatory notes③
	T of material handled per man-shift①	By type of operation						By type of cost			
		Drilling	Blasting	Loading	Haulage	General	Total	Labor	MEP②	Total operating	
1	299.7-299.7	0.052	0.067	0.050	0.088	0.064	0.321	0.115	0.206	0.326	3,5
2	76.9-206.8	0.020	0.020	0.039	0.157	0.089	0.323	0.079	0.246	0.330	3-5, 8-10, 15
3	142.2-426.7	0.024	0.036	0.030	0.076	0.031	0.197	0.072	0.125	0.197	2, 3, 5, 6, 8, 14, 16
4	NA	0.014	0.030	0.075	0.134	0.068	0.320	NA	NA	0.325	3-6, 8-10, 13-15
5	264.2	0.013	0.016	0.057	0.175	0.112	0.373	0.088	0.285	0.379	1-15
6	117.9-130.0	0.020	0.079	0.039	0.118	0.148	0.404	0.167	0.236	0.410	3,9
7	304.8	0.019	0.018	0.053	0.153	0.090	0.371	0.073	0.259	0.337	3-6, 8-10, 14, 15
8	165.6	0.013	0.020	0.048	0.084	0.049	0.214	0.044	0.169	0.217	2, 3, 5, 6, 8, 9, 14
9	61.8-80.3	NA	NA	NA	NA	NA	NA	NA	NA	NA	
11	47.8-211.3	0.025	0.012	0.042	0.131	0.053	0.263	NA	NA	0.267	3-5, 8-11, 14
13	116.0-201.0	0.010	0.010	0.050	0.135	0.070	0.275	0.146	0.129	0.279	2-6, 8, 10, 11, 14
15	62.2-255.4	0.036	0.024	0.056	0.116	0.053	0.286	0.139	0.148	0.291	2-14
16	33.5-67.1	0.017	0.030	0.049	0.085	0.073	0.255	0.069	0.183	0.259	1, 3-10, 11, 12
17	284.5	NA	NA	NA	NA	NA	NA	NA	NA	NA	3-8, 10
18	NA	0.011	0.021	0.031	0.136	0.049	0.248	NA	NA	0.252	3-6, 9, 11
19	165.6-435.8	NA	NA	NA	NA	NA	NA	NA	NA	NA	2-15, 17
20	556.8-556.8	NA	NA	NA	NA	NA	NA	NA	NA	NA	3, 5, 6, 8, 9
25	67.1-133.1	NA	NA	NA	NA	NA	NA	NA	NA	NA	3-6, 9-12
26	152.4	0.036	0.040	0.035	0.091	0.085	0.287	0.087	0.201	0.292	3-5, 8, 10, 11, 13-15
27	130.0	NA	NA	NA	NA	NA	NA	NA	NA	NA	3-6, 8, 10, 11, 14
28	117.4-141.5	NA	NA	NA	NA	NA	NA	NA	NA	NA	3, 8, 9, 14
29	20.6-16.4	0.102	0.155	0.164	0.307	0.501	1.224	0.502	0.722	1.244	2, 3, 5, 6, 8, 9
31	NA -NA	0.061	0.043	0.035	0.027	0.035	0.202	0.043	0.158	0.205	3-14
32	70.5-237.5	0.020	0.059	0.030	0.170	0.069	0.344	0.108	0.236	0.350	3, 5, 6, 8-12, 15
33	NA	0.018	0.033	0.035	0.091	0.041	0.219	0.041	0.177	0.222	2-11, 13-15

Continued 6-8

Mine	Productivity	Actual cost per t of material handled, $									Explanatory notes③
	T of material handled per man-shift①	By type of operation						By type of cost			
		Drilling	Blasting	Loading	Haulage	General	Total	Labor	MEP②	Total operating	
34	55.9-655.3	0.030	0.049	0.069	0.177	0.020	0.344	0.246	0.098	0.350	2-4, 6-9, 11-15
35	10.6-12.3	NA	NA	NA	NA	NA	NA	NA	NA	NA	3, 5, 6, 8, 10, 14
36	27.4-53.8	0.004	0.038	0.029	0.086	0.025	0.181	0.021	0.160	0.184	3-6, 8-10, 14, 15, 16

①First entry, tons per man-shift ore only; second entry, tons per man-shift, waste only.

② MEP. Materials, expense (other than labor) and power.

③ 1. Depreciation included.

General cost includes:

2. Area supervision.
3. Mine supervision
4. Employee benefits.
5. Overtime premium.
6. Mine-office expense.
7. Head-office expense.
8. Mine surveying.
9. Pumping.
10. Development drilling.
11. Payroll taxes.
12. State and local taxes.
13. Insurance.
14. Assaying.
15. Mine-plant depreciation.

Reporting years other than 1971:

16. 1970 to 1972.
17. 1970.

The ages and sizes of the operations included in the tables cover a wide range as do the equipment sizes, locations and lengths of haul roads, dumps, shifts worked per week, and the items included in "general mine cost".

Regardless of these differences in operations, the statistical and cost data contained in the tables provide usable reference material for developing data and costs for new properties, worthwhile comparisons between properties and basic information for improvements at existing properties.

6.3.3.5 Production Costs

Productivity and actual costs are given in Table 6-8. Strip ratios and unit operation percentage costs are presented on Table 6-9.

Table 6-9 Strip ratio and percentage cost (per t of material handled) **data for open-pit copper mines**

Mine	Strip ratio	By type of operation					Labor cost as a % of total	Explanatory notes①
		Drilling	Blasting	Loading	Haulage	General		
1	4.80	16.0	20.6	15.4	26.9	19.6	35.9	3,5
2	2.70	6.0	6.0	11.9	47.7	26.8	24.2	3-5, 8-10, 15
3	3.00	11.8	18.2	14.8	37.9	15.7	36.5	2, 3, 5, 6, 8, 14
4	2.80	4.2	9.1	23.0	41.2	20.9	NA	3-6, 8-10, 13-15, 16
5	2.68	3.3	4.1	15.1	46.3	29.6	23.5	1-15
6	1.13	4.8	19.2	9.6	28.7	36.0	41.5	3,9
7	4.40	5.5	5.3	15.7	45.3	26.6	22.0	3-6, 8-10, 14, 15
8	3.37	5.9	9.1	22.2	38.6	22.6	20.7	2, 3, 5, 6, 8, 9, 14
9	1.13	NA	NA	NA	NA	NA	NA	
10	0	5.9	31.5	11.8	13.8	35.4	15.0	3-10, 14
11	4.42	9.3	4.4	15.8	49.0	19.9	NA	3-5, 8-11, 14
12	0.74	NA	NA	NA	NA	NA	NA	3, 5, 9, 14
13	1.73	3.4	3.6	17.9	48.4	25.0	53.0	2-6, 8, 10, 11, 14
15	5.00	12.8	8.1	19.5	39.9	18.2	48.6	2-14
16	NA	6.5	11.4	19.8	32.7	28.1	28.2	1, 3-10, 11, 12
17	0.57	1.9	4.9	7.9	46.3	37.4	31.0	3-8, 10
18	2.30	4.3	8.2	12.5	53.9	19.5	NA	3-6, 9, 11
19	2.70	7.9	12.8	11.8	44.3	21.7	40.0	2-15, 17
20	1.00	5.7	19.9	16.6	35.8	20.4	33.0	3, 5, 6, 8, 9
21	1.83	NA	NA	NA	NA	NA	NA	3, 6, 8-10, 14, 15
22	4.28	NA	NA	NA	NA	NA	NA	3, 5, 6, 8-10, 13, 14
23	1.62	NA	NA	NA	NA	NA	NA	NA
25	2.33	14.2	10.9	17.6	38.6	17.1	21.0	3-6, 9-12
26	1.67	12.5	13.8	12.1	31.0	29.0	30.1	3-5, 8, 10, 11, 13-15
27	0	NA	NA	NA	NA	NA	NA	3-6, 8, 10, 11, 14
28	1.21	9.4	25.0	16.1	31.7	16.1	19.8	3, 8, 9, 14
29	0.80	8.3	12.4	13.2	24.3	40.3	41.0	2, 3, 5, 6, 8, 9
30	3.00	6.9	5.9	18.7	50.2	16.7	40.0	1-9, 11-15, 18
31	6.60	29.7	21.2	17.3	13.0	17.2	21.5	3-14
32	2.80	5.6	16.8	8.5	47.8	19.7	31.4	3, 5, 6, 8-12, 15

Continued 6-9

Mine	Strip ratio	By type of operation					Labor cost as a % of total	Explanatory notes①
		Drilling	Blasting	Loading	Haulage	General		
33	1.61	8.0	15.3	16.1	40.6	18.5	18.9	2-11, 13-15
34	3.50	8.5	14.1	19.7	50.6	5.6	71.4	2-4, 6-9, 11-15
35	4.10	NA	NA	NA	NA	NA	NA	3, 5, 6, 8, 10, 14
36	1.98	2.2	20.9	15.6	46.6	13.4	11.4	3-6, 8-10, 14, 15, 16

① 1. Depreciation included in each "Percentage Cost".
General cost includes:
2. Area supervision.
3. Mine supervision.
4. Employee benefits.
5. Overtime premium.
6. Mine-office expense.
7. Head-office expense.
8. Mine surveying.
9. Pumping.
10. Development drilling.
11. Payroll taxes.
12. State and local taxes.
13. Insurance.
14. Assaying.
15. Mine-plant depreciation.
Reporting Years Other Than 1971:
16. 1970 to 1972.
17. 1970.
18. 1968 to 1972.

The mining costs reported for the open-pit copper mines in the United States ranged from \$0.22 to \$0.37 per metric ton of material moved, with an average cost of \$0.29 per t. The Canadian mining costs ranged from \$0.20 to \$0.40, with an average of \$0.32. Reporting mines in the other countries ranged from \$0.20 to \$1.22, with an average of \$0.24.

The striping ratios for the reporting United States mine averaged 2.86 : 1; the Canadian mines, 2.19 : 1; and the other mines, 1.78 : 1.

Productivity at the United States mines averaged 250t of material per man-shift; at the Canadian mines, 244t; and the other mines, 135t.

The two lowest costs reported were about \$0.20 per metric ton of material mined. These two new properties started operations in late 1971, and both reported relatively short haulage distances and correspondingly low haulage costs, with indicated high availability of the new equipment. The highest cost reported was from a small isolated mine utilizing small-sized mining equipment in extremely wet conditions.

A comparison of the detailed cost breakdowns of properties in the United States with mines in other countries shows that the United States mines operate with generally lower drilling and blasting costs. The drilling costs average \$0.019 per metric ton of material at the United States mines compared to \$0.036 for the other reporting mines. The basting costs average \$0.019 per metric ton of material in the United States, compared to \$0.056 for the other properties.

Loading costs are about equal in the United States and Canada and average about \$0.02 per metric ton less than at reporting mines from other countries. Haulage cost in the United States average about \$0.137 per metric ton of material, compared to \$0.117 outside the United States. The general costs are in about the same range for all properties reporting.

The major single item causing higher haulage costs at the United States mines is their longer average haul distance, which averages 2.7km, compared to 1.7km at the other reporting properties.

The labor cost as a percentage of total mining costs is about 33.6% in the United States and Canada and only 25.9% at reporting mines in the other countries—a direct affect of the basic differences in wage rates of the different countries.

The availabilities and utilizations of the reporting mines are not absolutely comparable from the information reported, as many mines use different methods of calculating these figures, and operating schedules also differ and affect availabilities through having more or less time set aside for maintenance (Table 6-17).

6.3.3.6 Drilling and Blasting Practices

All the large open-pit copper mines use rotary drills for production drilling, whereas some of the smaller pits use percussion and down-the-hole units. Hole diameters vary from 102 to 330mm for the rotaries and from 51 to 178mm for the other drills. The most common rotary drill-hole size is 229mm with the 305 to 330mm hole size increasing in use (Table 6-10).

Table 6-10 Drilling practices at open-pit copper mines

Mine	Material	No. drills	Type drill	Penetration rate, m/h	Hole diameter, mm	Advance blasthole drilling, No. holes
1	All	1	Rotary	23	229	Nil
2	All	4	Rotary	13	229, 311, 349	NA
3	Ore	1	Rotary	26	251	25
	Waste	2	Rotary	26	251	50
4	Ore	NA	Rotary	24	251	20
	Waste	NA	Rotary	34	251	50
5	All	6	Rotary	15	229	NA
6	All	2	Rotary	12	229	NA
7	All	4	Rotary	14	229	NA
	All	1	Rotary	23	229	NA
	All	1	Rotary	13	229	NA
8	All	7	Rotary	15	276, 311	NA
9	Ore	1	Rotary	7	175, 229	NA
	Waste	1	Rotary	7	175, 229	NA

Continued 6-10

Mine	Material	No. drills	Type drill	Penetration rate, m/h	Hole diameter, mm	Advance blasthole drilling, No. holes
10	Ore	1	Rotary	9	251	245
11	All	NA	Rotary	13	171	30
12	All	2	Rotary	23	251	NA
13	All	2	Rotary	21	229	None
15	Ore	3	Rotary and	23	171	Minimal
	Waste		DTH	20	171	Minimal
16	Ore	5	Rotary	15	121	NA
	Waste			20	130	NA
17	Waste	5	Rotary	25	251	250
18	All	5	Rotary	24	229, 251	NA
19	All	1	Rotary	11	229	NA
	All	2	Rotary	15	229	NA
20	All	1	Rotary	18	311	NA
	All	1	Rotary	21	251	NA
21	Waste	7	Rotary	11	311	NA
	Leach	1	Rotary	15	311, 349	NA
	Ore	1	Rotary	17	349, 381	NA
	Flux	1	Rotary	7	200	NA
22	All	5	Rotary	27	311, 349	160
23	Waste	NA	Rotary	11	311	15
	Leach	NA	Rotary	14	311	15
	Ore	NA	Rotary	10, 14	311, 270	65
24	All	4	Rotary	NA	311	NA
25	All	4	Rotary	11	251	NA
26	All	3	Rotary	14	200	45
27	Ore	4	Percussion	16	64	NA
28	All,	4	Rotary			
	15-20, 000 psi,			15	251	100
	50, 000 psi			9	251	1, 000
29	All	2	Percussion	7	102	NA
30	Ore	2	Rotary	14	229	NA
	Waste	2	Rotary	21	229	NA
31	Ore	2	Percussion,	16	102	NA
	Waste	2	DTH	7	191	NA
32	All	2	Rotary	16	251, 311	NA
33	All	5	Rotary	14	251, 311	NA
34	All	4	Rotary	15	251	45
35	All	3	Percussion	20	76	NA
36	All	2	Rotary	12	159, 171	0

Note: 1psi = 6. 89kPa.

One operation reports using seismic-velocity techniques for determining blasting patterns and kilometers of "powder" per hole (Table 6-11). Throughout, the use of straight and metallized slurries is increasing.

Table 6-11 Blasting design at

Mine	Material	Pattern①	No. holes	Rows	Spacing, m	Burden, m	T blasted	Powder factor, kg/t	Explosives used②
1	All material	17F	150	8-10	5	2	101,600	0.348	ANFO
	Ore								
	Waste								
2	All	1H	20	1	8	671m^3	95,504	0.299	ANFO & MS
3	All	5F	40	2	7	18	81,280	0.134	ANFO & S
4	Overburden	16B	50	3	6	NA	NA	0.156	ANFO & S
	Waste	16B	40	3	5	NA	NA	0.156	ANFO & S
	Ore	16B	20	2	5×6	NA	NA	0.145	ANFO & S
5	All	1C	8-10	1	6	16	24-30,480	0.080	S & C
6	Ore	6F	14	4	10	12	26,416	0.202	ANFO & S
	Ore & waste	6F	14	4	11	14	26,416	0.172	ANFO & S
7	Seismic velocity:								
	0-2,400 fps	5F	40	4	11	23	122,936	0.170	ANFO
	2,500-4,000	5F	40	4	9	21	92,456	0.223	ANFO
	4,100-6,000	5F	40	4	7	16	56,896	0.362	ANFO
	6,100-15,000	5F	40	4	6	14	42,672	0.482	ANFO
	Ore								
	Waste								
8	Altered rock	1H	100	1	7	23	264,160	0.114	ANFO
	Porphyry	1H	100	1	5	23	243,840	0.117	ANFO
	Contact zone	1H	100	1	5	23	223,520	0.128	ANFO
9	Ore	14E	21	3	5	12	27,737	0.179	MS
	waste	14E	24	3	6	15	30,542	0.277	MS
10	Ore	4E	60-100	3-5	10	20	NA	0.295	S & MS
11	Overburden	15D	70	4	5	12	152,400	0.067	ANFO & S
	Leach	15D	70	4	4	11	142,240	0.067	ANFO & S
	Ore	15D	70	4	4	9	131,064	0.067	ANFO & S
12	Overburden	6H	100	4	8	19	NA	0.112	ANFO & S
	Rock	6H	100	4	7	18	203,200	0.201	ANFO & S
13	Soft	3A	8	1	8	34	40,640	0.075	ANFO & S
	Medium	3A	5	1	7	31	22,352	0.081	ANFO & S
	Hard	3A	6	1	6	34	30,480	0.089	ANFO & S
15	Ore	5C	24	2-3	4	11	11,582	0.165	ANFO & MS
	Waste	5C	16	1-2	4	9	4,877	0.223	ANFO & MS
	Waste	5C	16	1-2	4	11	6,909	0.156	ANFO & MS

open-pit copper mines

Water conds.	Delay time, hr		Secondary breakage		Crusher		Ore density, m^3/t	Bench ht., m	Subgrade drilling, m	Stem-ming, m
	Normal	Max.	%	Method	Type③	Prod. size				
Mod.	4	24						7.6	2.1	3.7
			5	Ball	J48×36	5.5	0.502	7.6	2.1	3.7
			1	Drill & AGY				7.6	2.1	3.7
								0.0	0.0	0.0
Mod.	2-3	24	2	Drill & GDY	J84×66	8.0	0.346	15.2	1.5	11.6
								0.0	0.0	0.0
Mod.	4	28	0.5	Drill & blast	G54	−9.0	0.321	12.2	2.7	7.0
Mod.	24	48	2	Drill & blast				10.1	2.1	6.1
Mod.	24.	48	2	Drill & blast				10.1	2.1	6.1
Mod.	24	48	1	Drill & blast	G54	−6.0	0.335	10.1	2.1	6.1
Mod.	10 min	NA	0.8	Drill & blast	G56	−8.0	0.508	12.2	1.5	7.3
Mod.	<24	72	5	Drill & blast	G42	−6.0	NA	10.7	1.5	4.6
Mod.	<24	72	0	None			NA	10.7	1.5	4.6
								0.0	0.0	0.0
None	48	72					0.360	10.7	2.7	7.6
None	48	72					0.360	10.7	2.7	7.6
None	48	72					0.360	10.7	2.7	7.6
None	48	72					0.360	10.7	2.7	7.6
			2.5	50% Drill & blast	J66×84	−6.0				
			5.0	50% plasters						
Varied	7days	3 wks	0-0.1	Drill & blast				14.9	1.8	9.1
Varied	7days	3 wks	0-0.1	Drill & blast	G60	10.0	0.340	14.9	1.8	9.1
Varied	7days	3 wks	0-0.1	Drill & blast			0.000	14.9	1.8	9.1
Mod.	60	720	3	Drop ball	J42×65	−6	0.195	11.0	1.8	5.2
Mod.	60	720	Nil	Drop ball			0.000	11.0	1.8	5.2
Mod.	NA	NA	0.6	Drill & blast	G54	8	0.340	14.9	2.4	1.5
Slight	1	24	1	Drill & blast				12.2	1.5	8.2
Slight	1	24	1	Drill & blast				12.2	1.5	8.2
Slight	1	24	1	Drill & blast	J42×40	6	0.349	12.2	1.5	8.2
Mod.	4	48	0					13.7	1.8	6.7
Mod.	4	48	0		G54	−7	0.346	13.7	1.8	6.7
Slight	0.25	0.5	1	Plasters, drill & bl.	G42	−6	0.349	15.2	2.4	7.9
Slight	0.25	0.5	1	Plasters, drill & bl.				15.2.	2.4	7.9
Slight	0.25	0.5	1	Plasters, drill & bl.				15.2	2.4	7.9
Slight	4-6	30	4	Plasters	G24	3	0.329	6.1	0.8	3.7
Slight	4-6	30	2	Plasters				6.1	0.8	3.7

Mine	Material	Pattern①	No. holes	Rows	Spacing, m	Burden, m	T blasted	Powder factor, kg/t	Explosives used②
16	Sandstone	NA	20	3	5	13	NA	NA	ANFO
17	Waste	6H	40-70	2-4	12	24	Up to 243, 840	0.107	S
18	All	4F	40	4	8	9	95, 504	0.134	ANFO & S
19	All	14C	21-30	2-3	3	8	NA	0.152	ANFO, S, MS
20	All	6E	60-80	2	10 × 9	20-17	195-20, 726	0.223	ANFO & S
21	Soft	3D	Varies	1	14	16	386, 080	0.380	ANFO & coal
	Medium	3D	Varies	1	12	16	243, 840	0.357	ANFO & coal
	Hard	3D	Varies	1	11	16	182, 880	0.357	ANFO & coal
22	All	2F	18	1	12	24	182, 880	0.080	ANFO & MS
23	Hard	2F	15	1	6	8	45, 720	0.447	ANFO & S
	Soft	2F	15	1	7	9	51, 308	0.335	ANFO & S
	Medium	2F	15	1	7	9	52, 578	0.391	ANFO & S
24	Ore	3F	15	1	10	16	55, 880	1.697	MS
25	All	14G	48	3-4	9	17	49-87, 376	0.250	ANFO & S
26	All	NA C	30	2-3	6 × 7	14-15	38, 608	0.143	S
27	Ore	NA	180	3	2	5	15, 240	0.201	ADY
28	– 10, 000 psi	14D	200	7-10	9	18	12, 903, 200	0.223	S
	15-20, 000 psi	14D	100	3-5	7	17	32, 004, 000	0.272	S
	50, 000 psi	14D	30	3-5	7	18	32, 613, 600	0.299	S
29	Limestone	14C	18	3	4	9	NA	0.089	ANFO
30	All	4E	25	2	6 × 5	7	50, 800	0.192	ANFO
31	Ore	9D	40	3-6	2-2.5	5-5.3	9, 144	0.357	ANFO, ADY
	Waste	9D	30	3-4	7-7.3	8-9	28, 448	0.201	& MS
32	Ore	5E&F	40	4	12	15	111, 638	0.201	ANFO & S
	Waste	5E&F	28	5	11	14	85, 049	0.179	ANFO & S
	Overburden	5E&F	20	4	13	17	59, 009	0.080	ANFO & S
33	Ore	17H	40	2	8	9	158, 496	0.143	ANFO & S
	Waste	17H	40	2	9	11	203, 200	0.143	ANFO & S
34	Ore	17E	90	5	10	12	182, 880	0.156	ANFO & MS
	Leach	17E	60	5	11	19	182, 880	0.134	ANFO & MS
	Waste	17E	90	5	10	12	182, 880	0.156	ANFO & MS
35	Ore	5B	240	6	1	8	1, 321	0.098	ANFO & GDY
36	Ore	14F	15	2	6	14	24, 384	0.107	ANFO & S
	Ore	14F	17	3	7	15	22, 352	0.098	ANFO & S
	Waste	14F	25	3	8	16	33, 528	0.067	ANFO & S

Note: 1fps = 0.3048m/s; 1psi = 6.89kPa.

① Patterns not to scale.

② Explosive used: ANFO, ammonium nitrate and fuel oil; GDY, gelatin dynamite; ADY, ammonia dynamite; S, slurry; C, carbonite; AGY, ammonium gelatin dynamite; MS, metallized slurry.

③ G, gyratory crusher; J, jaw crusher.

Continued 6-11

Water conds.	Delay time, hr		Secondary breakage		Crusher		Ore density, m^3/t	Bench ht., m	Subgrade drilling, m	Stemming, m
	Normal	Max.	%	Method	Type[3]	Prod. size				
Slight	4-6	30	2	Plasters				6.1	0.8	3.7
None	0.5	2	1	Blast	G NA	−4	0.410	NA	NA	NA
Slight	NA	NA	Nominal	Blast	G54	−8	0.357	15.2	1.5	8.5
Mod.	4	72	−1	Drill & blast	G54	−8	0.352	12.2	3.7	8.2
Varied	2	24	1	Drill & blast	G48	−6	0.349	12.2	3.0	7.6
Mod.	72	720	−1	Drill & blast	G60	7-8	0.335	15.2	2.4	7.9
Slight	1	48						15.2	4.6	10.7
Slight	1	48	3	Block-hole				15.2	4.6	10.7
Slight	1	48	10	Block-hole	G72	5	0.349	15.2	4.6	10.7
Slight	6	24	0.25	Drill & blast	G60	−6.5	0.343	15.2	4.6	10.4
Mod.	4	24	15	Drill & blast	G48	6	0.349	15.2	4.3	9.8
Mod.	4	24	10	Drill & blast				15.2	4.3	9.8
Mod.	4	24	10	Drill & blast				15.2	4.3	9.8
Mod.	2	20	−1	Drill & blast	G54	6	0.349	12.2	3.0	9.8
Mod.	1 wk	3 wk	0.5-1.5	Drop Ball	G54	−7	0.2-0.3	12.2	1.5	4.6
Mod.	1	3	2.5	Drill & blast	J66 × 84	8	0.357	10.1	2.4	5.5
Mod.	12	24	3	Drill & blast	J42 × 48	7	0.234	NA	NA	NA
Slight	96	144	2	Drill & blast				12.5	2.1	4.6
Slight	96	144	2	Drill & blast	G54	7	0.279	12.5	2.1	4.6
Slight	96	144	2	Drill & blast				12.5	2.1	4.6
Slight	2	NA	5	Drop ball & bl.	Impact	4	0.349	6.1	1.2	2.4
Slight	4	24	Slight	Drill & blast	G54	−6	0.326	12.2	3.7	6.7
Mod.	8-24	200	10	Drill & blast	J48 × 60	4-6	0.307	12.2	1.2	2.7
Mod.	8-24	200	1.5	Drill & blast				12.2	1.8	4.3
Mod.	12	48	1	Drill & blast	G54	−9	0.335	15.2	2.4	9.8
to	12	48	1	Drill & blast				15.2	2.4	9.8
Severe	12	48	0					15.2	2.4	7.3
Mod.	4	24	Nil	Plasters	G58	−8	0.349	15.2	2.1	9.1
Mod.	4	24	Nil	Plasters				15.2	2.1	9.1
Slight	48	168	0.5	Drill & blast	G84	6	0.335	12.2	1.5	5.5
Slight	48	168	0	Drill & blast				12.2	1.5	5.5
Slight	48	168	0	Drill & blast				12.2	1.5	5.5
Slight	0.5	1	10	Mud capping	J	4	NA	8.5	1.8	1.2
Mod.	3	36	2.5	Mud capping	J	7	0.343	10.1	2.1	4.9
Mod.	3	36	2.5	Mud capping				10.1	2.1	4.9
Severe	3	36	2.0	Mud capping				10.1	2.1	4.9

Table 6-11 (Continued)

LEGEND

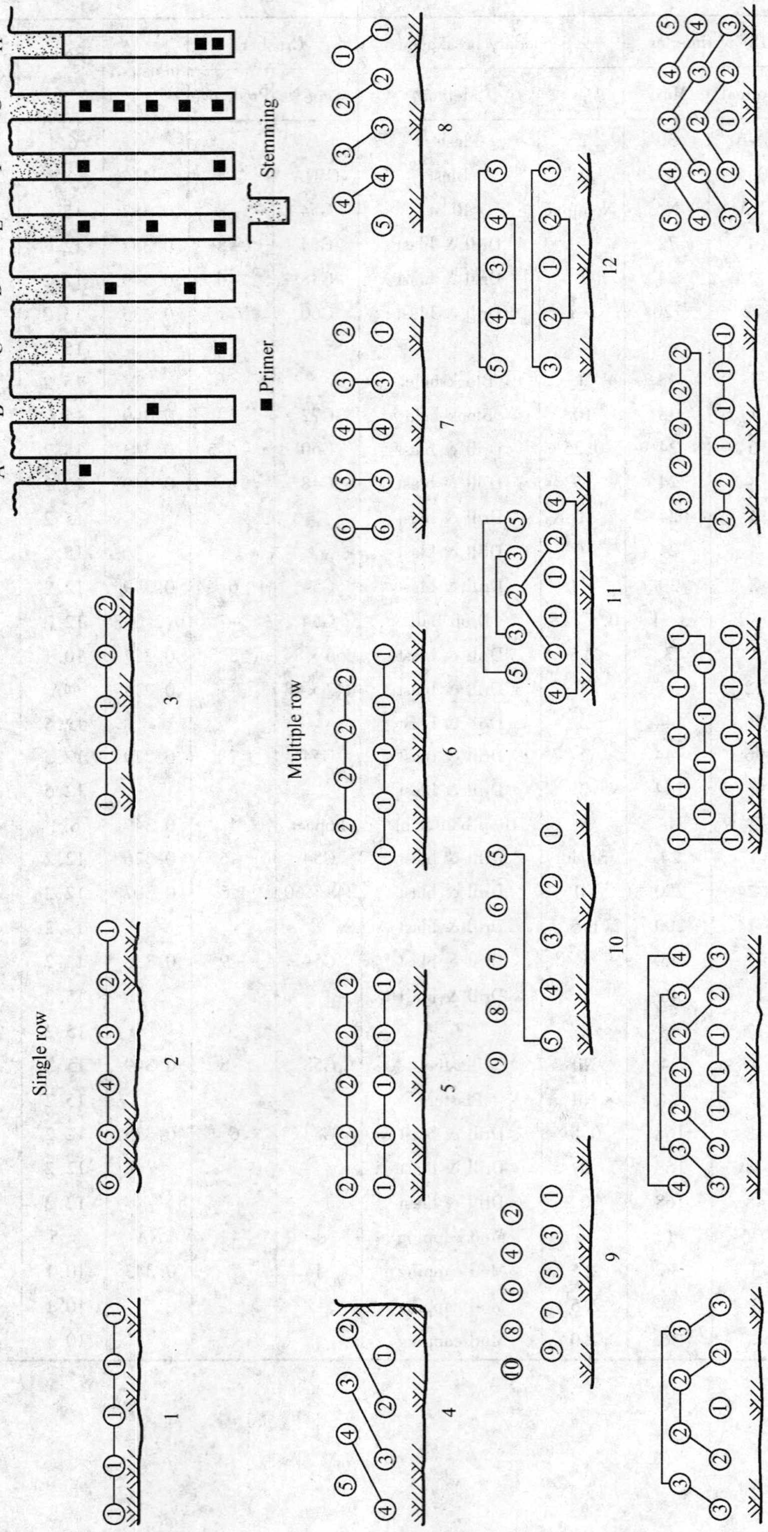

The general trend in rock breaking is toward larger-diameter hole sizes and increased hole spacing. The use of bulk mixing trucks is increasing and more mines are reporting the use of higher-density powders and stronger boosters.

6.3.3.7 Loading and Haulage Practice

The trend of replacing small shovels, front-end loaders and trucks with larger units is continuing and can be expected to accelerate as a means of overcoming higher direct labor rates and their related indirect incurred costs.

The use of front-end loaders to replace or supplement small shovels is increasing, and they are being used increasingly where high-speed mobility, operating flexibility and selective ore mining are controlling factors (Table 6-12).

Table 6-12 Loading practices at open-pit copper mines

Mine	Material	No. shovels	Size, m^3	T per shovel shift①	Operating and maintenance cost, $ per t
1	All	1	3.8	5,070	0.040
	All	1	3.4	4,531	0.067
	All	1	1.9	2,499	0.052
2	All	7	4.6	5,618-4,999	0.049
	All	1	6.1	7,295-7,112	0.039
	All	2	11.5	14,427-13,117	0.020
	Reload ore	2	4.6	6,604	0.030
3	All	4	7.6	11,176-11,176	0.030
4	All	4	4.2	4,775-4,775	0.080
	All	Loaders, 3	9.2	6,096-7,112	0.080
5	All	3	4.6	5,568-4,600	0.065
	All	5	5.4	5,779-5,608	0.065
	All	3	11.5	9,388-11,030	0.043
6	All	2	3.8	3,251-3,556	0.039
7	All	1	3.4	3,489-3,962	0.086
	All	3	3.8	3,656-4,550	0.080
	All	1	4.6	4,767-4,787	0.077
	All	2	6.9	8,740-8,491	0.048
	All	1	9.9	12,598-12,598	0.042
	All	1	11.5	16,662-16,662	0.034
8	All②	5	6.9	5,486	Avg. cost,
	Waste	5	6.9	7,620	0.048
	Ore	1	6.9	7,010	
	Waste	1	3.4	NA	
9	All	2	3.8	2,540-2,540	NA
10	Ore	1	5.7	3,658	NA
11	All	1	3.8	4,518	Avg. cost,
	All	1	3.8	4,573	0.038
	All	1	4.6	6,599	
	All	1	6.9	10,408	

Continued 6-12

Mine	Material	No. shovels	Size, m^3	T per shovel shift①	Operating and maintenance cost, $ per t
12	All	3	10.7	13,208-13,208	NA
13	All	3	3.8	5,588-7,894	0.051
	All	2	4.6	5,402-4,800	0.058
	All	1	6.1	7,435-9,624	0.059
	All	1	7.6	10,925-13,116	0.028
	All	Loader, 1	7.6	5,475-6,168	0.059
	All	Loader, 1	6.9	3,709-3,726	0.179
	All	Loader, 1	4.6	4,675-4,465	0.090
14	Ore	Scraper, 6	36t	1,853	0.265
15	All	Loader, 7	4.6	2,709-3,304	0.057
16	All	3	3.4	Avg. t,	Avg. cost,
	All	Loader, 2	3.8	4,313	0.051
17	Waste	6	13	12,588	NA
18	All	4	11.5	13,208	0.029
	All	4	6.1	6,604	0.040
19	All	4	3.8	4,953	NA
	All	1	5.4	6,690	NA
20	All	2	8.4	7,925-7,925	NA
21	All	10	4.6	4,064	NA
	All	1	5.4	4,064	NA
	All	4	6.9	6,299	NA
	All	3	11.5	11,735	NA
22	All	6	7.6	11,766-11,689	NA
	All	3	11.5	14,254-16,036	NA
23	All	4	5.4	7,021-6,109	NA
	All	2	6.9	9,346-7,932	NA
24	All	14	4.6-6.9	3,911-5,015	NA
25	All	2	4.6	3,048-4,572	NA
	All	3	3.1	2,438-3,048	NA
	All	Loader, 3	7.6	2,845-2,845	NA
26	All	3	4.6	6,096-6,096	Avg.
	Waste & leach	Loaders, 2	4.2	4,572	0.036
27	Ore	Loaders, 3	4.6	2,438	0.086
28	All	2	3.4	4,948	NA
	All	5	4.6	7,366	NA
	Waste	1	9.2	17,221	NA
29	All	3	1.9	1,422-1,422	0.164
30	Ore	7	6.9	9,144	NA
	Rock			10,668	NA
	Waste			11,684	NA
31	Waste	1	4.6	4,381	0.027
	Waste	Loader, 1	7.6	4,665	0.035
	Ore	Loader, 1	4.6	2,959	0.035

Continued 6-12

Mine	Material	No. shovels	Size, m^3	T per shovel shift①	Operating and maintenance cost, $ per t
32	All material	3	11.5	7,478-7,478	0.028
	Over burden	1	11.5	11,938	0.015
	Ore stockpile	1	5	NA	NA
33	All	5	11.5	15,697-15,697	0.036
34	All	3	11.5	13,716	0.020
	All	1	6.1	7,112	0.039
	All	2	3.4	3,556	0.049
35	Ore	1	1.1	279	0.121
	Waste	1	1.1	167	0.192
	All	Loader, 2	1.1	303-120	NA
36	Ore	3	3.2	2,032	0.027
	Waste	4	3.2	1,524	0.030
	Waste	1	4.6	7,925	0.034

① First entry, ore only; second entry, waste only; center entry, average;

② Rail haulage only.

All of the large, new, open-pit copper mines reported the use of shovels ranging in size from 7.6 to $13m^3$. One leach copper operation reported the use of scrapers for loading and hauling all material.

In comparing loading costs and productivity per operating shift, the 7.6 to $13m^3$ capacity shovels averaged 12,395t per shift at a cost of $0.031 per t of material, the shovels in the 1.5 to 7.6m^3 capacity range averaged 5,486t per shift at a cost of $0.057, and the 3.8 to $9m^3$ front-end loaders averaged 3,861t per shift at a cost of $0.070 per t of material loaded.

As the haulage distance has increased, especially in the United States mines, the horsepower of existing trucks has been increased, paralleling the purchase of new, larger-size and horsepower replacement trucks (Table 6-13). This trend is forecasted to continue at many of the reporting mines as their pits get deeper and the haul distances increase.

Another trend reported is the movement away from all-rail haulage. Those mines that started with rail haulage now are increasing their use of truck haulage for waste removal.

Only one operation reports the use of an input crusher. The crushed ore is dropped through an ore pass and then hoisted to the surface. Another new operation reports the use of an aerial tram system to transport ore to the concentrator (Table 6-14).

6.3.3.8 Supporting Operations

Diamond drilling is the general practice for exploration and development work (Table 6-15). Existing pit equipment ordinarily is used for pioneering, with only one mine reporting the use of a contractor and dragline on such work.

Truck-haulage cost accounts for about 39% of the total mining cost and, since truck costs are strongly affected by the condition of the haul roads, greater attention is being given to haul-road maintenance. There is an increasing trend away from the use of run-of-the-pit material for road construction, with many mines now reporting the use of crushed rock and borrow material for improved road construction and maintenance (Table 6-16). Some mines reported the trial use of different type road binders to attempt to further improve haul-road conditions.

Table 6-13 Truck-haulage practice at open-pit copper mines

Mine	Material	No. trucks	Haulage-truck nominal rating		Haulage profile			Operating data			
			Capacity (0.9072t), t	Engine, kW	Haul, m	Max. grade, %	Elevation, (gained +, lost −), m	T per truck shift	Availability, %	Utilization, %	Operating and maintenance cost, $ per t
1	All	6	36	298	914	10.0	+76	975	71.0[1]	NA	0.080
	All	2	36	298	914	10.0	+76	975	76.0[1]	NA	0.094
2	All	10	66	522	2,368	7.0	+73	1,097	76.0[2]	44.1[4]	0.148
	All	6	66	522	2,670	7.0	+88	1,046	60.1[2]	52.2[4]	0.187
	All	6	86	522	3,054	7.0	+104	1,300	81.9[2]	80.9[4]	0.128
	All	3	86	522	2,707	7.0	+88	1,382	62.4[2]	61.3[4]	0.138
	All	3	86	522	3,176	7.0	+113	1,311	84.0[2]	83.2[4]	0.108
	All	1	102	746	3,200	7.0	+119	1,453	68.8[2]	67.7[4]	0.157
	All	3	102	746	2,954	7.0	+107	1,524	76.3[2]	75.6[4]	0.138
	All	3	102	746	2,905	7.0	+107	1,636	73.1[2]	72.3[4]	0.098
3	All	15	102	746	1,524	10.0	+122	2,743	82.0[3]	73.0[5]	0.076
4	Waste	7	51	447	1,829	12.0	−	1,727	60.0[3]	53.0[5]	0.098
	Ore	3	51	447	1,829	12.0	+	1,118	60.0[3]	53.0[5]	0.167
	Ore	2	51	447	1,829	12.0	+	1,118	65.0[3]	53.0[5]	0.167
5	All	17	66	522	2,591	10.0	+98	1,321	60.0[2]	54.0[4]	0.151
	All	11	86	522	2,591	10.0	+98	1,727	61.0[2]	60.0[4]	0.113
	All	5	76	447	2,591	10.0	+98	1,524	76.0[2]	73.0[4]	0.149
	All	4	102	746	2,591	10.0	+98	2,032	69.0[2]	68.0[4]	0.099
	All	3	122	746	2,591	10.0	+98	2,438	84.0[2]	83.0[4]	0.052
6	Ore	4	51	474	1,097	10.0	−32	987	83.8[3]	71.8[5]	0.118
	Waste	4	51	474	1,768	10.0	−30	987	83.8[3]	71.8[5]	0.118
7	All	15	66	474	NA	NA	NA	1,120	57.0[2]	39.0[4]	0.200
	All	7	76	522	NA	NA	NA	1,347	60.0[2]	41.0[4]	0.181
	All	20	122	746				2,259	68.0[2]	56.0[4]	0.113
	Ore				3,111	4.8	−38				
	Waste				2,832	9.6	+134				
8	Waste	12	86	522	2,499	8.0	+24- −16	1,953	95.5[1]	94.3[6]	0.065
	Ore	2	76	522	1,554	6.5	+14- −13	1,787	91.5[1]	90.8[6]	0.075
	All	30	38	317	1,372	6.5	+20- −19	868	90.7[1]	83.2[6]	0.091

Continued 6-13

Mine	Material	No. trucks	Haulage-truck nominal rating: Capacity (0.9072t), t	Haulage-truck nominal rating: Engine, kW	Haulage profile: Haul, m	Haulage profile: Max. grade, %	Haulage profile: Elevation, (gained +, lost −), m	Operating data: T per truck shift	Operating data: Availability, %	Operating data: Utilization, %	Operating data: Operating and maintenance cost, $ per t
9	All	9	46	388	NA	8.5	+76	914	80.3[2]	NA	NA
	All	4	51	453	NA	8.5	+76	1,016	80.3[2]	NA	NA
	All	5	71	340	9,656	NA	+61	711	73.1[2]	NA	NA
10	Ore	4	66	436	610	8	+15- −50	1,204	25.0, NA	8.0, NA	NA
11	All	18	27	283	1,746	7	NA	Avg.	94.6	18.4	Avg.
	All	15	53	447	1,875	7	NA	1,626	74.4[3]	67.4[5]	0.130
	All	1	86	597	1,682	7	NA		78.4[3]	73.1[5]	0.000
12	All	13	102	738	1,067	10	+76	1,778	NA[2]	77.0[4]	NA
	All	1	152	1,193	1,067	10	+76	2,667	NA[2]	NA[4]	NA
13	Ore	29	41	391	2,037	7.0	Variable	851	47.4[2]	39.5[4]	0.202
	Waste				2,037	7.0	Variable	1,075			
	Ore	3	86	522	2,037	7.0	Variable	1,778	44.5[2]	31.9[4]	0.083
	Waste				2,037	7.0	Variable	2,600			
	Ore	5	36	309	2,037	7.0	Variable	805	50.5[2]	40.1[4]	0.132
	Waste				2,037	7.0	Variable	1,012			
	Ore	1	76	522	2,037	7.0	Variable	1,602	26.1[2]	18.6[4]	0.078
	Waste				2,037	7.0	Variable	2,297			
	Ore	5	76	597	2,037	7.0	Variable	1,709	64.5[2]	56.9[4]	0.055
	Waste				2,037	7.0	Variable	2,378			
14	Ore	Scrapers, 6	36	298	640	6.0	+46	1,853	70.0[2]	67.0[4]	0.265
15	Ore	18	36	309	2,469	10.0	+61	734	73.0[2]	69.0[4]	0.174
	Waste				914	10.0	+91	1,289			0.100
16	Waste	20	51	474	1,829	8.0	+61	1,121	75.0[1]	75.0[6]	Avg.
	Ore	6	51	474	3,658	8.0	+61	169	66.7[1]	66.0[6]	0.085
17	Waste	41	102	746	2,286	−10.0	−168	1,748	70.9[2]	68.0[4]	NA
18	All	10	122	716	4,183	8.0	+213	1,219	81.0[3]	76.0[5]	0.133
	All	23	102	716	4,183	8.0	+213	1,219	81.0[3]	76.0[5]	0.137
	All	17	86	522	4,183	8.0	+213	914	80.0[3]	76.0[5]	0.137
19	Waste	5	66	474	1,561	8.0	+116	1,509	78.8[3]	32.2[5]	NA
	Waste	7	76	522	1,078	8.0	+116	1,879	78.4[3]	72.8[5]	NA
	Ore	10	76	474	5,794	8.0	+82	906	73.3[3]	55.4[5]	NA
20	All	12	102	522	1,067	8.0	NA	2,032	80.0[2]	65.0[4]	NA
21	Waste	4	66	391	1,067	8.0	+38	2,032	NA	NA	NA
22	All	32	86	522	2,807	8.0	−21	2,277	80.0[2]	37.0[4]	NA
	All	22	102	746	2,292	8.0	−19	3,094	86.0[2]	81.0[4]	NA
23	All	28	66	522	3,175	10.0	+91	1,250	81.0[2]	NA	NA
	All	10	66	597	3,175	10.0	+91	1,250	81.0[2]	NA	NA

Continued 6-13

Mine	Material	No. trucks	Haulage-truck nominal rating		Haulage profile			Operating data			
			Capacity (0.9072t), t	Engine, kW	Haul, m	Max. grade, %	Elevation, (gained +, lost −), m	T per truck shift	Availability, %	Utilization, %	Operating and maintenance cost, $ per t
24	Waste	11	67	522	787	8.0	+27	2,213	73.0[2]	55.0[4]	NA
25	All	20	51	466	1,524	10.0	−148	890	70.0[2]	55.0[4]	NA
26	All	27	36	309	1,829	−8.0	−37	762	85.0[3]	76.0[5]	0.091
27	Ore	3	51	522	701	−15.0	−105	1,626	93.0[1]	93.0[6]	0.118
28	All	11	66	447	1,951	9.0	+61	1,534	Avg.	Avg.	NA
	All	8	66	500	1,951	9.0	+61	1,534	92.5[2]	92.5[4]	NA
	All	5	66	403	1,951	8.0	+61	1,534			NA
	Waste	2	102	641	1,951	8.0	+46	2,977			NA
	Waste	8	102	746	1,951	9.0	+46	2,977			NA
29	All	13	15	114	1,219	10.0	+73	249	72.0[2]	NA	0.302
30	Ore	7	76	597	2,743	8.0	+61	1,311	75.0[3]	60.0[5]	NA
	Waste	18	76	522	4,572	9.0	+61	1,626	75.0[3]	60.0[5]	NA
31	Ore	3	36	309	1,067	−7.6	−6	9,144	89.0	71.0	0.018
	Waste				1,067	7.6	+37	9,144	89.0	71.0	0.018
	Waste	8	51	NA	1,067	7.6	+37	1,219	85.0	60.0	0.009
32	Ore	6	122	746	1,829	10.0	+59	Avg.	Avg.		
	Waste	6	122	746	1,524	10.0	+84	1,321	75.7[2]	73.4[4]	0.173
	Over-Burden	4	122	746	1,372	10.0	+30				
33	Ore	12	122	746	1,768	10.0	+46	2,540	76.9[3]	66.5[5]	Avg.
	Waste	18	122	746	2,880	10.0	+91	1,676	76.9[3]	66.5[5]	0.090
	Waste	5	152	1,007	2,880	10.0	+91	3,048	80.2[3]	69.8[5]	
34	All	12	102	746	1,219	8.0	+91	2,032	75.0[5]	75.0[6]	0.069
	All	4	122	895	1,219	8.0	+91	2,438	75.0[5]	75.0[6]	0.059
	All	5	51	447	1,219	8.0	+91	1,016	75.0[5]	75.0[6]	0.079
	All	6	86	522	1,219	8.0	+91	1,829	75.0[5]	75.0[6]	0.079
	All	2	30	224	1,219	8.0	+91	1,016	50.0[5]	10.0[6]	0.118
35	Ore	7	13	NA	NA	4.0	NA	201	59.0[5]	46.0[6]	0.172
	Waste							135			0.295
36	Waste	5	20	177	1,814	8.0	+35	305	85.0[5]	75.0[6]	0.429
	Ore	11	36	298	1,585	8.0	+85	914	85.0[5]	85.0[6]	0.085
	Waste	24	36	298	1,737	8.0	+60	813	85.0[5]	85.0[6]	0.089

Notes: 1. Availability equals percentage time truck actually in production on a 7-hr-per-shift, 1-, 2-, or 3-shift-per-day basis.
2. Availability equals percentage time truck actually available on an 8-hr-per-shift, 3-shift-per-day basis.
3. Availability equals percentage time truck actually available on a 7.5-hr-per-shift, 1-, 2- or 3-shift-per day basis.
4. Utilization equals percentage time truck actually in production on an 8-hr-per-shift, 3-shift-per-day basis.
5. Utilization equals percentage time truck actually in production on a 7.5-hr-per-shift, 1-, 2-, or 3-shift-per-day basis.
6. Utilization equals percentage time truck actually in production on a 7-hr-per-shift, 1-, 2-, or 3-shift-per-day basis.

Table 6-14 Other haulage practices at open-pit copper mines

Mine	Material	Type haulage	Locomotive		Rail-car data		Rail profile			T hauled per train-shift	Operating and maintenance cost, $ per t
			Deadweight, t	Hp	Capacity, 907kg, t	Avg. cars per train	Avg. haul, m	Max. grade, %	Elevation (gained +, lost −), m		
5	Ore	Railroad	152	2,300	102	40	7,711	1.5	−49	8,636	0.055
7	Ore	Railroad	142	1,500	71	48	28,968	1.6	−14	4,279	0.375
8	Waste	Railroad	124	1,775	73	11	4,522	4.0	−18	3,212	0.059
	Ore	Railroad	138	1,850	73	11	4,522	4.0	−47	3,212	0.059
10	Ore	Skip			17.6, skip				+140	1,324 per hr.	0.021
11	Ore	Conveyor		200	914mm belt		549	—	+88	330 per hr.	Avg.
	Ore	Conveyor		200	914mm belt		137	—	+37	1,016 per hr.	0.020
15	Ore	Conveyor		75	1,219mm belt		192	—	+67	508 per hr.	0.019
21	Waste	Railroad	127	1,800	85	11	7,315	3.0	−23	2,358	NA
	Ore	Railroad	127	1,800	85	11	7,315	3.0	+15	2,358	NA
24	All	Railroad	138	1,750	61	7	8,167	3.0	+168	1,637	NA
			138	2,250	85	8	8,167	3.0	+168	1,637	NA
27	Ore	Aerial tram			10.7, skip				−792	427 per hr.	0.010
36	Ore	Conveyor		400	1,219mm belt		316		+68	1,524 per hr.	0.021

Table 6-15 Exploration and development drilling and pioneering at open-pit copper mines

Mine	Exploration and development drilling					Pioneering		
	Drill units			Hole spacing in m		Shovel size, m^3	Truck size, t	Performed by
	Type	Core size, mm	m per shift	Exploration	Development			
1	None					3	36	Company
2	Diamond	BX	4.6-5.5	120	61	12	66, 86, 102	Company
3	Wireline	BX	7.6	91	30	8	102	Company
4	Diamond	NQ	12	120	61	NA	NA	NA
	Rotary	143	24-120	overburden only				
5	Diamond	NX, BX, MC	22.0	180	91	5, 5	66, 76	Company
6	NA	NA	NA	NA	NA	4	53	Company
7	Rotary	229	42.0	120	61	5	66	Company
	Diamond	HQ	15.9					
8	None					3	NA	Company
9	NA	NA	NA	NA	NA	4 & loaders	46, 53	Company
10	NA	NA	NA	NA	NA	6 dragline	None	Contractor
11	Wireline	AX	5.5	Variable	76	4	30	Company
12	None					11	102	Company
13	Churn	152-406	NA	Variable	Variable	Loaders, 5, 8	36, 41	Company
	Diamond	N-A	NA					
15	None			Random	61-120	Loader, 5	36	Company
16	Rotary	NA	1.8	76-152	30-46	3	53	Company
17	Diamond	NA	NA	120	NA	13	102	Company
18	Diamond	NX	9	120	61	NA	NA	NA
19	Rotary	229	78	46-61	NA	4	66, 76	Company
		229	107					
20	Na	NA	NA	NA	NA	8	36, 102	Company
21	Diamond	BX, NX	8	NA	NA	5	66, 86	Company
22	Diamond	NX	11	120	61	8 & 12	86, 102	Company
26	Rotary	187, 200	6	None	18 ×23	None	None	None
30	Diamond	NX	NA	180	88	7	76	Company
31	Diamond	BX, AX	12	38-152	Slight	NA	NA	Company
32	Diamond	NQ	18-21	120	76	12	122	Company
34	Diamond	DQ	11	180	30-61	12	102	Company
	Diamond	AQ	12					
35	Rotary	30	4	24-30	9-18	NA	NA	NA
	Rotary	30	5	24-30	9-18			
	Rotary	48	7	79-101	18-24			
36	Diamond	HQ, NQ, BQ	2	500	50	3	36	Contractor
	Rotary	171	91		7			

Table 6-16 General service equipment, road-maintenance practice and snow frequency at open-pit copper mines

Mine	Service equipment—No. units						Road maintenance					Snow removal
	Dozers	Graders	Pickup trucks	Passenger cars	Man buses	Mechanical & electrical veh.	Type surface	Cost per t of material moved, $	Type of road binder	Sprinkling frequency	L per km per sprinkling①	Days per avg. year②
1	2	1	17	1	0	4	Crushed waste	0.30	None	2 per shift	5,683 W	40
2	13	5	42	0	9	37	Waste	0.01	None	2-4 per shift	11,365 W	0
											11,365 NaCl	3
3	5	2	7	0	1	7	Gravel	0.20	None	12 per shift	8,524 W	35
4	4	2	20	1	1	4	Crushed waste	NA	None	NA	NA W	90
5	9	4	25	1	4	10	Rock	0.014	None	Max	505,743 per mo. W	0
6	2	1	6	0	1	2	Gravel	NA	Lignin Sulfonate	2 per shift	8,524 W	45
7	8	4	45	5	0	21	Waste	0.01	Sulfonate	3 per shift	42,619 W	6
8	25	9	60	7	26	129	Waste	NA	None	6 per shift	5,683 W	0
9	2	1	5	0	0	3	Gravel	NA	NA	NA	NA	25
10	NA	NA	1	4	0	1	Asphalt	NA	Salt	None	—	90
11	11	3	27	NA	0	18	Dirt	NA	None	6 per shift	7,103 W	0
12	4	2	20	1	1	7	Waste	NA	None	NA	14,206 W	30
13	4	2	9	0	2	8	Waste	0.019	None	Max	NA W	0
14	4	2	5	0	0	2	NA	NA	NA	NA	NA W	0
15	3	2	10	0	2	3	Waste	0.015	None	6 per shift	14,206 W	0
16	4	1	6	4	0	4	Alluvium	—	None	Max	14,206 W	0
17	18	5	36	4	7	11	Crushed stone	NA	None	4 per shift	1,762 W	0
18	5	3	13	0	0	7	Rock	0.008	None	6 per shift	5,683 W	0
19	7	1	6	2	1	11	Dirt	8%	None	7 per shift	NA W	0
20	4	2	2	0	2	5	NA	NA	Nil	6 per shift	14,206 W	120
21	19	6	39	1	0	50	Dirt	NA	SSKK	1 per shift	15,627 W	0
22	8	5	14	1	4	39	Crushed rock	NA	MgCl	5 per shift	NA W	6
23	8	3	25	1	1	26	Limestone & Slag	NA	None	13 per shift	9,802 W	0
24	14	4	39	1	3	31	NA	NA	NA	NA	NA	0
25	7	2	5	1	0	NA	Crushed rock	0.009	None	NA	NA	80
26	5	2	18	0	Con-tractor	8	Compacted slag	NA	None	Max	150 gpm	0
27	4	1	4	0	0	2	Crushed stone	NA	None	0	0	0
28	11	4	31	3	0	10	Crushed rock	NA	None	NA	5,683 W	0
29	2	1	2	0	0	1	Crushed stone	NA	None	2 per shift	NA W	0
30	8	2	4	0	4	6	Compacted gravel	NA	Cohesex	8 per shift	2,841 W	0
31	2	1	3	3	1	7	Rock	NA	None	11 per shift	13,638 W	0
32	7	3	10	4	3	7	Waste	0.34	None	NA	NA	NA
33	10	4	30	1	4	20	NA	0.02	None	5 per shift	8,524 W	2
34	7	2	4	0	2	12	Crushed rock	0.01	None	NA	NA	NA
35	4	2	5	0	3	2	NA	NA	NA	NA	NA	0
36	12	4	2	0	2	4	Waste rock	NA	None	2 per shift	6,819	0

① Letter following numerical entry indicates type sprinkling: W, water; C, chemical.
② Snow removal generally is accomplished with existing pit equipment. Only Companies 9 and 12 reported maintaining additional snow-removal equipment.

Table 6-17 Operating schedules for open-pit copper mines

Mine	Scheduled shifts per day		Scheduled days per week		Scheduled hr per operating shift
	Ore	Waste	Ore	Waste	
1	3	3	5	5	7
2	3	3	7	7	8
3	2	3	5	5	7.5
4	3	3	7	7	7.5
5	3	3	6	7	8
6	2	3	5	5	7.5
7	3	3	6	7	8
8	3	3	6	6	7
9	2	NA	6	NA	8
10	2	0	5	0	8
11	2	2	5	5	8
12	3	3	7	7	8
13	3	3	7	7	8
14	0	3	0	7	8
15	3	3	7	7	8
16	2	3	6	6	7
17	3	3	7	7	8
18	3	3	7	7	8
19	2	3	7	7	8
20	3	3	7	7	8
21	3	3	6	5.5	8
22	3	3	6	6	8
23	2	2	5.5	5.5	8
24	3	3	6	6	8
25	3	3	7	7	8
26	2	2	6	6	7
27	2	2	5	5	7.5
28	3	3	6	6	8
29	2	2	5	5	8.5
30	3	3	7	7	7.5
31	1	2	5	5.5	9
32	3	3	7	7	8
33	3	3	7	7	7.5
34	2	3	7	7	8
35	2	3	7	7	6.5
36	2	2	7	7	10

Professional Terms

mining method selection 采矿方法选择
underground mining method 地下采矿方法
open-stope mining 空场法
filled stopes method 充填法
caving method 崩落法
pillar 矿柱
shrinkage stoping 留矿法
sublevel caving 分段崩落法
surface mining (open pit mining) 露天开采
open-pit mine 露天矿
quarry 采石场
level 中段
sublevel 分段
raise 天井

stope preparation 采准
drawpoint 放矿点
dilution 贫化
LHD unit (load-haul-dump unit) 铲运机
drilling machine 钻机
drill rig 凿岩台车
drill bit 钻头
drilling-and-blasting 凿岩爆破，打眼放炮
support 支护
down-the-hole drill 潜孔钻机
front-end loader 前端装载机
shovel 机械铲
electric shovel 电铲
backhoe 反铲
dragline 索斗铲
clamshell 抓铲
scraper 铲运机

Exercises

6-1 影响采矿方法选择的因素有哪些？

6-2 请画出留矿法示意图。

6-3 简述分段崩落法的优缺点。

6-4 简述充填法的优缺点。

6-5 表 6-8 中的数据项"NA"代表什么？

6-6 1 (美)短吨(st) 等于多少公吨(t)？

References

[1] Ivan A. Given. SME Mining Engineering Handbook, Volumes 1 and 2, Society of Mining Engineers of the American Institute of Mining, Metallurgical, and Petroleum Engineers, Inc. New York, N. Y., USA, 1973.

[2] Julius Rubinstein, Lev A. Barsky. Non-ferrous Metal Deposits (On the example of Former USSR). TAYLOR AND FRANCIS, 2002.

[3] http://chinadataonline.org/member/hygk/hygkmshow.asp? code = 09.

[4] http://www.chengxinml.com/mlid/free/tjnj/html/F0614e.htm.

[5] Metalworld Correspondent. China's non-ferrous metal industry - its trend and impact for future outlook [J]. Metalworld. 26(2008).

[6] 占丰林，蔡美峰．高硫矿山高温采场的成因、危害及防治措施[J]．矿业研究与开发，2006，26(1)：71 ~ 73.

[7] 占丰林，杨建永，饶运章，黄文钿．东乡铜矿Ⅶ号难采矿体采场顶板控制技术[J]．南方冶金学院学报，1996，17(4)：237 ~ 241.

[8] 易德新．缓倾斜极厚难采矿体采矿方法的试验研究[J]．有色金属：矿山部分，1996，79(2)：9 ~ 14.

[9] 林睦曾．岩石热物理学及其工程应用[M]．重庆：重庆大学出版社，1991.

[10] 冯卫国．积分变换[M]. 2 版．上海：上海交通大学出版社，2000.

[11] 王志方．浅论砂浆长锚索支护技术[J]．矿山技术，1990，5：9 ~ 12.

[12] 占丰林，蔡美峰．地温梯度对地下矿山地应力计算的影响[J]．矿业研究与开发，2006，26(3)：24，25，48.

[13] 占丰林，刘湘萍，李春超．高温后裂隙岩体锚杆锚固相似模拟试验研究[J]．采矿技术，2007，7(1)：48，49，55.

[14] 占丰林，蔡美峰．高温对采场顶板锚索加固系统的影响[J]．辽宁工程技术大学学报，2007，26(4)：524 ~ 526.

[15] Eugene P. Pfleider, George B. Clark, Howard L. Hartman, and Adolph Soderberg. Surface Mining, the American Institute of Mining, Metallurgical, and Petroleum Engineers, Inc. New York, N. Y., USA, 1968.

[16] 占丰林．高温环境下岩体锚固机理——“胀锚拱理论”[J]．辽宁工程技术大学学报，2008，27(5)：686 ~ 688.

[17] F. L. Zhan, M. F. Cai. Rock heat-transfer model of high-temperature stopes and its solving process. Proceedings of The International Young Scholars' Symposium on Rock Mechanics, 2008, Beijing, China.

[18] 占丰林．高温环境下岩体锚固机理研究[D]．北京：北京科技大学，2006.

冶金工业出版社部分图书推荐

书　　名	定价(元)
采矿手册(第1卷~第7卷)	927.00
现代金属矿床开采技术	260.00
采矿工程师手册	395.00
中国冶金百科全书·采矿	180.00
现代矿山企业安全控制创新理论与支撑体系	75.00
充填采矿技术与应用	55.00
矿山废料胶结充填	42.00
矿井风流流动与控制	30.00
露天矿山台阶中深孔爆破开采技术	25.00
采矿知识问答	35.00
采矿概论	28.00
地下采矿技术	36.00
难采矿体新型采矿法	88.00
采矿学	39.80
中国爆破新技术Ⅱ	200.00
工程爆破实用手册(第2版)	60.00
炸药化学与制造	59.00
中国典型爆破工程与技术	260.00
金属矿山尾矿综合利用与资源化	16.00
常用有色金属资源开发与加工	88.00
非金属矿加工技术与应用	119.00
冶金矿山地质技术管理手册	58.00
非高炉炼铁工艺与理论	39.00
焦炉煤气净化操作技术	30.00
大学英语写作与修辞	36.00
现代英语词汇学	18.00
专业英语教程——材料成型与控制工程(英文版)	26.00
英汉冶金工业词典(修订本)	138.00
汉英英汉连续铸钢词典	65.00
当代英语教学论	20.00
英语教学理论探讨与实践应用	22.00
实用求职英语	25.00
冶金专业英语(本科教材)	28.00